Forging a Common Bond

New Perspectives on the History of the South

Florida A&M University, Tallahassee
Florida Atlantic University, Boca Raton
Florida Gulf Coast University, Ft. Myers
Florida International University, Miami
Florida State University, Tallahassee
University Of Central Florida, Orlando
University Of Florida, Gainesville
University Of North Florida, Jacksonville
University Of South Florida, Tampa
University Of West Florida, Pensacola

The New Perspectives on the History of the South series provides a forum for the best new scholarship on the ever changing South. The series emphasizes books that focus on the region's social, economic, cultural, and intellectual history, especially those with an interdisciplinary emphasis. The goal is to ask fresh questions about the South and to provide stimulating answers, arguments that in turn add to the ongoing dialogue of historical analysis about that ever fascinating land, the South.

Forging a Common Bond

Labor and Environmental Activism during the BASF Lockout

Timothy J. Minchin

Foreword by John David Smith, Series Editor

University Press of Florida

Gainesville · Tallahassee · Tampa · Boca Raton
Pensacola · Orlando · Miami · Jacksonville · Ft. Myers

Library of Congress Cataloging-in-Publication Data
Minchin, Timothy J.
Forging a common bond: labor and environmental activism during the BASF lockout /
Timothy J. Minchin.
p. cm. — (New perspectives on the history of the South)
Includes bibliographical references and index.
ISBN 0-8130-2580-X (cloth: alk. paper)
1. Chemical industry—Louisiana—History. 2. Badische Anilin- & Soda-Fabrik—
History. 3. Labor unions—Louisiana—History. 4. Oil, Chemical, and Atomic Workers
International Union—History. 5. Collective labor agreements—Louisiana—History.
6. Environmentalism. I. Title. II. Series.
HD6515.C432 O356 2003
331.892'861'0097631909048—dc21   2002028932

The University Press of Florida is the scholarly publishing agency for the State
University System of Florida, comprising Florida A&M University, Florida Atlantic
University, Florida Gulf Coast University, Florida International University, Florida
State University, University of Central Florida, University of Florida, University
of North Florida, University of South Florida, and University of West Florida.

University Press of Florida
15 Northwest 15th Street
Gainesville, FL 32611–2079
http://www.upf.com

Contents

Figures

Foreword

For all its alleged homogeneity and conservatism, the American South has always been a land of diversity, adaptability, and (to outsiders) surprising change. Though its rural, agriculturally based, and slave-driven antebellum economy retarded industrialization, labor organization nevertheless took root in the post-Reconstruction South. By the 1880s the Knights of Labor had organized around fifty thousand southern laborers, both white and African-American. Soon after, several American Federation of Labor national trade unions began in the South, including, for example, the International Brotherhood of Boilermakers, Iron Shipbuilders, and Helpers (1888) and the National Association of Postal Supervisors (1908). Though trade union membership expanded during World War I, it declined during the 1920s, fell even more behind the rest of the nation during the Depression, and thus the region became a fertile field for labor organizers following World War II.

In *Forging a Common Bond*, Timothy J. Minchin, who teaches at the University of St. Andrews, Scotland, presents a carefully researched and nuanced case study that suggests the complexity of southern labor-management relations in the last quarter of the twentieth century. Drawing upon an unusual and broad range of sources, including union and company records as well as detailed oral history interviews, he chronicles the fascinating story of a landmark labor dispute and the unusual alliance between a small union, Local 4-620 of the Oil, Chemical, and Atomic Workers' International Union (OCAW), and environmentalists in Ascension Parish, Louisiana, against the German corporate giant Badische Anilin und Soda-Fabrik (BASF).

In June 1984 BASF locked out its 370 hourly workers at its Louisiana plant, located near the small town of Geismar. After BASF introduced cheaper and more malleable replacement workers, the OCAW launched a corporate campaign against BASF and took its cause directly to the public both in the United States and in Germany. In doing so, the union mobilized an unlikely set of allies—Baptists, Roman Catholics, local community groups, German chemical workers, and conservationists, including the Sierra Club, Greenpeace, and members of the German Green Party. Ultimately this coalition exposed the giant multinational corporation's dismal environmental, health, and safety record and forced BASF to the bargaining table. When, in December 1989, the smoke of industrial battle had cleared, the underdog union had defeated the corporate giant. The OCAW had taken "a principled stand against concessionary bargaining."

Minchin's fast-paced, pathbreaking study suggests that workers in Geismar were no more passive or complacent to organize and protest than those elsewhere in the United States. The sustained corporate campaign that workers in Geismar launched against BASF appealed effectively to their local community to challenge the practices of a foreign-owned company that proved insensitive to local customs. As Minchin shows so well, the interests of laborers and environmentalists are not necessarily antithetical. In Geismar these two groups collaborated on tactics, thus advancing both groups' causes. This partnership continued after the lockout, and Local 4-620 became widely recognized for its willingness to work with environmentalists.

Readers will find Minchin's study a balanced, gripping, and important analysis of industrial conflict in the post–World War II South, a period that historian Charles P. Roland has aptly termed "the improbable era." The outcome of the labor-management battle in Geismar certainly *was* "improbable." During the five-year struggle, southern unionists built broad coalitions, depended on unity between themselves and German workers, and benefited from BASF's sensitivity to negative publicity on both sides of the Atlantic. As Minchin notes, "The union's survival in Geismar contradicts the images of defeat and decline often associated with organized labor in the 1980s and 1990s."

Forging a Common Bond is an important contribution to the burgeoning field of southern labor history, one that underscores the diversity and richness of the region's past and present, and the unpredictability of its future.

John David Smith
Series Editor

Acknowledgments

As I researched and wrote this book, I was assisted by a great many people from both sides of the BASF dispute. I owe a great debt to Leeann Anderson at the headquarters of PACE (Paper, Allied-Industrial, Chemical, and Engery Workers' Union) in Nashville. Leeann helped me tremendously, especially by ensuring that all relevant files from the lockout were made available to me and that I had a place to look at them. In PACE's Washington, D.C., office, Richard Miller opened many of his files relating to the lockout and also spared me a lot of his time. Richard Leonard was likewise generous with his time throughout the project.

I am also grateful for all the assistance I received from BASF. At BASF's headquarters, John Kirkman tracked down former managers involved in the dispute and helped me arrange interviews with them. I would like to thank former BASF executives Bill Jenkins, Richard Donaldson, and Les Story, all of whom were particularly helpful.

I was aided a great deal by many other people in the United States. I am especially obliged to my friends Chuck and Yvonne Spence for providing a home away from home. In Louisiana, Willie Fontenot assisted me in fixing up interviews and shared with me his vast knowledge of the environmental movement in Louisiana. The staff of the Louisiana Labor–Neighbor Project put up with me working in their offices, while several BASF workers helped me in arranging interviews. I am especially indebted to Bobby Schneider, Roger Arnold, Duke King, and Roy Fink. All the members of PACE Local 4-620 made me feel very welcome when I was in Louisiana, and I am very grateful to them for that. I would also like to thank the staff at the Louisiana Environmental Action Network (LEAN) for allowing me open access to all of their files and providing me a place in which to read them.

John Clark, my friend and colleague at St. Andrews University, was particularly supportive, while former colleague Tony McElligott helped by recommending reading in German history. I would also like to thank the research committee of the School of History at St. Andrews for supplying funding for this project. Both Tony Badger and Bob Zieger gave valuable advice and guidance throughout the project, and I am also grateful to Alan Draper and Gary Fink for offering helpful comments on an earlier draft. I would moreover like to express my appreciation to the British Academy, who awarded me a research grant that funded most of the research for this project.

I would also like to acknowledge the staff of the University Press of Florida, particularly editor in chief Meredith Morris-Babb, who has dealt with the manuscript in a very efficient way and has been encouraging and helpful throughout. My final thanks are to my family, particularly my parents and my wife, Olga, for all their support and love.

Introduction

On December 15, 1989, representatives from the Oil, Chemical, and Atomic Workers' International Union (OCAW) and the BASF corporation reached agreement on a new contract for the company's large plant in Geismar, Louisiana. The settlement brought to an end one of the longest labor disputes in recent labor history, a conflict dating back to June 1984, when the German chemical giant had locked out 370 workers at the plant after the two sides had been unable to reach agreement on a new contract. More than three years later, BASF lifted the lockout for the plant's production workers. Over 100 maintenance workers had not been allowed to return, however, and the OCAW continued an active corporate campaign against BASF until all of its members had been given the opportunity to go back to work. Achieving this goal took more than five and a half years and involved both the union and the environmental groups working together to put pressure on the giant German chemical-maker. This environmental alliance was the central feature of the BASF dispute and the main reason for its eventual resolution.

The union's alliance with environmentalists involved its joining forces with groups as varied as the Sierra Club, Greenpeace, and the German Green Party. This alliance does not fit with the popular notion that environmentalism and labor are fundamentally at odds. This image is based on a number of high-profile conflicts between unions and environmentalists, conflicts that occurred primarily because labor feared that cleaning up the environment would threaten their jobs. In the 1970s and 1980s, for example, organized labor and environmentalists clashed over the Alaska Pipeline, the Tellico Dam, clean air laws, and the expansion of the Redwoods National Park. More recently, the battle between West Coast loggers and environmental activists fighting to save the northern spotted owl has continued the image. Radical environmentalists sabotaged bulldozers and removed survey stakes, prompting verbal and physical retaliation from the loggers. Throughout the 1980s and 1990s, many loggers drove around displaying bumper stickers that proudly read "I Love Spotted Owls—Fried," "Save a Logger, Eat an Owl," and "If you object to logging, try using plastic toilet paper."[1]

The union's alliance with environmentalists led to a permanent transformation in the way that the workers involved viewed industrial-pollution issues. Throughout the 1990s, the local union continued its environmental work, funding a variety of projects designed to make industry more accountable to the public interest. OCAW Local 4–620 became, in the words of environmental historian Jim Schwab, "one of the most environmentally progressive and innovative union locals in the nation."[2]

In many respects, the BASF dispute was representative of broad trends in labor relations. Like most labor disputes of the 1980s, it was the company and not the union that sought major concessions when the two sat down to negotiate. In the summer of 1981, President Ronald Reagan fired 13,000 striking air traffic controllers, a move that encouraged many employers to take a hard line at the bargaining table.[3] If workers resisted, many companies were willing to take strikes and continue operating.[4] At the same time, plant closings across the country led to many union jobs being lost. Union influence declined precipitously throughout the decade, with membership falling from 23.2 percent of the nonagricultural workforce in 1980 to 16.1 percent in 1990.[5]

When workers did not strike, many companies responded by locking them out of their jobs, a tactic that very few corporations had used before the 1980s. In October 1986, for example, the *New York Times* noted that "More and more companies are adopting the lockout tactic to force further concessions from give-back-weary unions." Major lockouts, the *Times* noted, were now affecting "tough, once-untouchable unions" such as the United Automobile Workers and the United Steelworkers. In the 1980s, a wide variety of workers were locked out of their jobs, including ballet dancers in New York, shipbuilders in Washington, machinists in Florida, and beer-truck drivers in New Jersey. Among the biggest examples of the decade were national disputes affecting 22,000 workers at USX Corporation (formerly U.S. Steel), and 12,000 employees of John Deere.[6]

The BASF lockout was, however, the longest of all.[7] In what the German chemical giant's executives themselves termed "a textbook case conflict," the union responded to the lockout by launching a coordinated corporate campaign.[8] Since 1981, many unions have turned to such campaigns, which aim to bring pressure against businesses at the highest levels of the corporate hierarchy, in an effort to try and exert some leverage against companies that no longer fear strikes. "The option is to fight in new ways the company isn't used to, or accept more losses," explained Joseph Uehlein, special projects director for the AFL-CIO's Industrial Union Department. The track record for corporate campaigns is, however, mixed. Union efforts against Phelps Dodge, International Paper, Louisiana Pacific, and General Dynamics are considered failures, but there were also more successful campaigns against Litton Industries, Beverly Enterprises, and Campbell Soup Company.[9]

When OCAW leaders decided to launch a corporate campaign against BASF, they knew their task was not going to be straightforward. In many ways, the BASF lockout was a classic David and Goliath story, pitting a small union against a giant multinational corporation that employed more workers than the OCAW's total membership. The union pressed ahead, however, because it felt that important principles were threatened. Like many U.S. unions, the OCAW had lost power in the early 1980s as employers had become increasingly successful at securing concessions at the bargaining table. Employers had stripped away the wages and benefits of OCAW members and had been able to persuade many workers that they no longer needed union representation. The union saw the BASF dispute as "a line in the sand," their opportunity to take a stand against concessionary bargaining and send a message to industry that they would resist it. "It is our intention," wrote OCAW strategist Richard Leonard, "and that of the Industrial Union Department of the AFL-CIO, who is working with us on the project, to make an example of BASF—to show Corporate America what can happen when an employer puts all decency aside and engages in a frontal assault on its union workers." The union's leadership committed a huge amount of resources to the campaign because they felt that their credibility as an organization was at stake in the dispute.[10]

This commitment ultimately paid off. Unlike many unions in the 1980s, OCAW Local 4–620 was able to fight back effectively against the company and resist many of their concessionary demands. The OCAW's corporate campaign secured nationwide press attention. The union bypassed the traditional forums of labor disputes—the negotiating table and the company gates—to take its cause directly to the public. In the process, the workers mobilized a broad coalition of civic and community groups to help them fight the company. The union subjected BASF's environmental record to particular scrutiny, which made the company uncomfortable and finally brought it back to the bargaining table.[11]

1

The Trade-Off

A visitor who traveled along the Mississippi River through Louisiana at the start of the twentieth century would have seen an agricultural landscape that had changed little for more than two centuries. On the former slave plantations that flanked the river, African-American laborers worked land that was still largely owned by the descendants of slave-owners. Louisiana's economy was, as the *Baton Rouge Morning Advocate* put it, "as sleepy as the state's moss-draped bayous."[1]

This picture began to be changed by successful oil exploration in the early 1900s, which was followed by the construction of a refinery in Baton Rouge. The petrochemical industry in Louisiana took off during the economic boom set in motion by American entry into World War II, and by 1956 it directly employed over 87,000 people. In the 1950s and 1960s, chemical companies moved into southern Louisiana in force, buying up large tracts of land next to the Mississippi River and transforming the area's economy. At its peak in 1982, Louisiana's petrochemical industry employed 165,000 people, and industrial taxes accounted for one out of every three tax dollars collected by the state. More than one hundred chemical plants were packed along the eighty-mile stretch of the Mississippi River between Baton Rouge and New Orleans. Together, they produced over 20 percent of America's petrochemicals, turning out products as varied as herbicides, plastics, soap, PVC piping, shower curtains, lipstick, and kitchen wrap. Louisiana had become the third-largest chemical-producing state, outstripped only by New Jersey and Texas.[2]

The petrochemical industry was attracted to Louisiana primarily by its abundance of natural resources. "God did not give Louisiana white beaches," explained Dan Borne, the president of the Louisiana Chemical Association (LCA). "He gave Florida all the beautiful, white, pristine beaches. He gave Colorado mountains. He gave Mississippi beautiful women; they've had three Miss Americas come from Mississippi. What he gave Louisiana was dead dinosaurs. He gave us a Jurassic Park of raw materials; oil and gas basically, that come from fossils that are located far beneath the earth, and so from that oil and gas base has come this huge industry, which is based on the basic raw materials of oil and gas. . . . So Louisiana got this incredible gift of all these raw materials."[3]

Located just south of Baton Rouge, Ascension Parish became the heart of Louisiana's burgeoning chemical industry. By the 1980s, thirteen companies sprawled around the bend in the Mississippi River near Geismar, a small, predominantly African-American community. Companies found that they could buy large tracts of land right next to the river, which was large enough to handle oceangoing vessels that brought raw materials in and took finished products out. "Ascension Parish," noted an LCA publication, "reaped the harvest of location, land, and water. On the Mississippi river, Ascension is accessible by deep draft vessel, and it has been an active area for industrial growth." From 1969 to 1994, employment in Louisiana grew by 65 percent, but in Ascension Parish it expanded by 217 percent.[4]

State politicians, keen to attract good-paying jobs, encouraged industrial development. As state governments did in other southern states, Louisiana's leaders courted industry after World War II because they felt that good-paying industrial jobs would strengthen the economy and lift their state out of poverty. In the 1960s, Governor John McKeithen, described by Borne as "a very forward-looking industrial developer," traveled around the world and invited large chemical manufacturers to come to Louisiana, telling them about the state's natural advantages. Companies such as ICI and Shell were lured by McKeithen's tax exemptions. Edwin Edwards continued McKeithen's policies, embarking on trips to seek more industry. A four-time governor of the state, Edwards is the dominant figure in post–World War II Louisiana politics. First elected governor in 1971, Edwards had an ability to bring industry to the state, which attracted support from a broad coalition of voters, including many unionized workers. While in office, he consistently defended industrial tax-exemption laws and was frequently pictured welcoming new chemical companies to Louisiana.[5]

The press also welcomed industry. "Industrialization means both work and upgraded jobs with higher per capita income," noted the *New Orleans Times-Picayune* in . 1973. Louisiana's biggest newspaper described the chemical industry as a "bulwark" of the state's economy, noting that "for every extra dollar earned by chemical workers, the result to the state's economy is $1.53 in personal income."[6] Local newspapers celebrated the arrival of plants into their communities, feeling that industry would provide a better, more prosperous life for local people. Industry, for its part, used the press to sell its message that Louisiana would prosper with industrial jobs. In May 1958, for example, the Louisiana Chambers of Commerce sponsored "Business-Industry Appreciation Week," with advertisements in local newspapers that confidently proclaimed, "Louisiana is Strong Thanks to Business and Industry. . . . We're on road to a better way of life and we can thank our business and industry for that."[7]

Many plants were built on the sites of former plantations, which offered cheap and unincorporated land. The availability of land was a major attraction for Wyandotte Chemical Company when it decided to locate in Ascension Parish. In 1958, Wyandotte purchased over 1,200 acres of riverfront land that had formerly been three separate slave plantations and opened a plant that made components for automobile antifreeze (ethylene glycol). The purchase gave the company much greater opportunities to expand than it had at its cramped home site in Wyandotte, Michigan, located on the edge of Detroit. "Wyandotte was not a good location for a chemical plant," recalled David Gilin, a salaried worker who transferred from Wyandotte to Geismar. "It's in the middle of a very populated area. High labor costs up there, high land costs." Shortly after opening the original plant, the company purchased a further 1,200 acres and built several new plants, including a large chlorine facility that opened in 1968.[8]

The opening of the Geismar plants marked the first time in Wyandotte's seventy-year history that the company had expanded operations outside of its Michigan base. Executives felt a considerable degree of culture shock when they arrived in Louisiana from the Midwest. "When the plant was started in 'fifty-eight," recalled one manager, "all of the management came from Michigan, and these people were coming into an area they just couldn't believe, they just didn't know. Baton Rouge was the closest civilization." Although other companies later joined them, Wyandotte was one of the first chemical firms to locate in predominantly rural Ascension Parish. Shocked by the lack of infrastructure in the area, the company's executives lived in Baton Rouge and hired almost of all of their first workers from the city, insisting that there were not enough qualified skilled workers available in Ascension Parish.[9]

Wyandotte was a family-owned company, and workers recalled that managers came and interviewed each potential job applicant individually at their home. Once offered a position, workers regarded it as a job for life. Esnard Gremillion, who later became chairperson of the union, be-

1. Esnard Gremillion. (Courtesy PACE International Union)

gan working at the plant when it first opened. "When we first started out there," he recalled, "the company officials that did the hiring visited you in your home and talked to your family and this kind of stuff and said once you work for us you have a job for a lifetime. That's the way people went to the job, thinking they were, you know, good for what they said."[10]

Gremillion, widely known as "Gremmy," remembered that workers felt a strong sense of identification with the company, and there were few barriers between hourly and salaried workers: "There wasn't no fence around the perimeter like they've got now. We'd go out there and hunt and fish and do whatever we wanted to, and when it first started it was just like a family. We sent everybody Christmas cards and when somebody was born and all that kind of stuff. There wasn't that much difference between supervisors and the working people really, on a social basis. We all knew one another pretty well, we knew their families." There was, as another worker put it, "harmony" between hourly and salaried workers.[11]

Shortly after the opening of the Geismar plant, it was organized by the OCAW, a union created by the 1955 merger of the United Gas, Coke, and Chemical Workers and the Oil Workers' International Union. The merger was an attempt to create a stronger voice for workers in the oil and chemical industries, especially as the two predecessor unions had overlapping jurisdictions and had often negotiated with the same companies. Although the OCAW did have more chemical workers signed up than any other union, it still struggled to secure a strong presence in the industry; by the early 1980s, for example, it represented only 20 percent of organized chemical workers. Unionism in the chemical industry remained fragmented, with workers belonging to a wide variety of unions, including the United Steelworkers and the International Chemical Workers' Union. With a multiplicity of unions and reasonably low levels of representation, there was no coordinated bargaining in the chemical industry; unions had to negotiate with companies individually rather than being able to secure a pattern-setting agreement with several major producers at once. Chemical workers also made up only a small portion of the OCAW's members, many of whom worked in the oil and nuclear power industries.[12]

Along Louisiana's chemical corridor, most plants were also nonunion. Ernie Rousselle, the OCAW's international union representative in the area, traveled the length of the corridor in the 1960s and 1970s as he serviced a handful of local unions. The OCAW had made some headway in organizing the first plants to build on the corridor in the 1950s and 1960s, but after this companies that came in usually kept their wages higher than those of the union plants, thereby helping to forestall further unionization. Although Local 4-620's members had made some efforts to organize workers in other local plants, they all insisted that these workers felt that they did not need a union to receive high wages. In addition, as union presence in the corridor gradually declined, workers in nonunion plants who could see the benefits of having an independent voice in the workplace were often afraid of being fired if they supported organized labor.[13]

Unions' weak position in the corridor was reflected across the United States, as organized labor struggled to sign up the majority of chemical workers. Du Pont, the largest company in the industry, was widely viewed as "the citadel of antiunionism in the chemical industry."[14] Between 1945 and 1981, unions held 235 representation elections at Du Pont plants but could only claim 5 percent of the company's 66,000 employees as members. The company emphasized its family-worker traditions and competitive salaries to explain why workers remained unorganized, while unions blamed determined corporate opposition and the difficulties of organizing a company as large and spread out as the giant chemical-maker.[15]

Unlike most chemical companies, however, Wyandotte Chemical Company did not vigorously oppose unionization. All of Wyandotte's plants in Michigan were organized, and executives had no experience of running a nonunion plant. As a consequence, management insisted that they had invited the union into Geismar, a view supported by some workers. "Wyandotte was the family-owned business and they had to have a union," reflected Bobby Schneider.[16] While some workers claimed that Wyandotte did not welcome the union as much as it later claimed, after OCAW Local 4-620 secured its first charter in early 1959 the company worked constructively with the local union, and contracts were renewed without major disputes throughout the 1960s. The union was strongly supported by the workers, and it negotiated a contract that they were proud of, with generous overtime provisions and strict seniority protection. Union members used the grievance procedure frequently when they felt that the company had violated these rules. Many protested, in particular, when they felt that overtime had not been assigned according to the contract, or when they perceived that workers had been assigned to jobs outside their classification.[17]

Local 4-620's members repeatedly stressed the noneconomic benefits of unionism, particularly their rights to file grievances and to be protected by the seniority system against arbitrary layoff or dismissal. "I realized the difference between working union and nonunion," explained Carey Hawkins, who had worked in both environments. "You worked just as hard, you got paid maybe less. You had just as many problems as the guys across the street but there was more respect, I think, an equality thing. You know your seniority, the overtime distribution. Those things was like fair and equal, and I liked that idea because you can't help but work and you see favoritism and nepotism."[18]

As Hawkins explained, this emphasis on noneconomic benefits provided a way for workers to value their union even though it did not necessarily deliver higher wages than those paid to nonunion workers in the area. Workers also genuinely felt strongly about the way that the union protected them against arbitrary treatment by the company, insisting that these benefits had to be experienced to be fully appreciated. Workers viewed the seniority system as "sacred," the key vehicle that they could use to protect themselves against arbitrary and unfair company actions. "One of the key things was the seniority," noted Roy Fink. "The rule of seniority is the first one hired is the last one fired. . . . and without the seniority, they could do with you as they please." Bobby Schneider, like most workers, felt that the seniority system more than compensated for the fact that nonunion workers in other chemical plants often received slightly higher wages: "We were a little below everybody else on wages but we had the security of our seniority and our layoff clauses that all of the nonunion didn't have, and the nonunion plants get the money just a little bit above to try and have an incentive to keep from going union."[19]

The first dispute between the two sides occurred in 1970, when workers struck the plant for nine months. As in 1984, the two sides clashed over seniority, with the union unwilling to consider any dilution of restrictions on job assignments. Workers also sought improvements in their wage and benefit package. Although the strike lasted for nine months, the company made no attempt to hire replacements, running the plant with salaried workers instead. In March 1971, the strike was settled on terms favorable to the union, with the company granting "substantial" wage increases, together with increases in health insurance and retirement benefits. The union also secured a clause in the contract that gave workers more rights to refuse to perform jobs that they felt threatened their personal safety. The company was unable to secure any substantial change in the contract's seniority provisions.[20]

Although it was rather isolated, the Geismar union was, as Rousselle recalled, always a strong local. Until 1976, when the Louisiana state legislature passed a law prohibiting compulsory union membership, a clause in the contract required all workers to belong to Local 4-620. Even after the closed shop was made illegal, close to 100 percent of workers belonged to the union. In addition, the contract contained a clause to reinstate the closed shop if organized labor could succeed in repealing the right-to-work law, which it was not able to do. The establishment of the union had clearly been aided by the lack of opposition from Wyandotte; while many chemical companies in Louisiana vigorously resisted organized labor, the Wyandotte workers benefited because the company did not strongly oppose their local union.[21]

The local union also had long-serving, dedicated leaders who helped to keep it strong. From 1966 until 1979, when he became a vice-president of the OCAW, the local was serviced by Rousselle, who was well-known and respected by all the workers. Like many of the Geismar workers, Rousselle came from an Acadian French (Cajun) background and was a practicing Roman Catholic. He had grown up near New Orleans and had worked for a local chemical plant for fourteen years before becoming an OCAW representative. Rousselle was widely known as "Big Ernie," and BASF managers recognized the close bond that existed between him and Local 4-620's members. "Ernie was one of them," noted BASF executive

2. Ernie Rousselle *(by stop sign)*. In the background are locked-out workers.
(Courtesy PACE International Union)

Bill Jenkins. "They were Cajuns and Ernie had been with them a long
time. . . . the members trusted him." Jenkins, who was the company's director of industrial relations, added that he respected Rousselle's intellect.
After serving a two-year term as an OCAW vice-president, Rousselle
returned to Louisiana in 1981 and serviced the local union for the remain-
der of the decade.[22]

Local 4-620 also possessed a cadre of strong rank-and-file leaders,
many of whom worked in the maintenance department. Since the maintenance workers moved around the plant to perform their jobs, they became
well-known to other union members. At the time of the 1984 negotiations, the two top leadership positions in the local union were held by
John Daigle and Esnard Gremillion, who were both maintenance workers. Originally composed exclusively of workers from BASF, Local 4-620
had, by the time of the lockout, expanded to include workers at the neighboring Vulcan Chemical plant. The Vulcan plant opened in 1968 and was
organized by the OCAW the following year. Although Vulcan workers
also belonged to OCAW Local 4-620, they bargained separately and had
their own contract. Daigle, an experienced and well-respected union

leader, served as the president for both groups while Gremillion was chairperson for the BASF group within the local. Gremillion won his position in July 1983 by defeating Harold "Doc" Nickens. BASF managers regarded Gremillion as a more militant and outspoken union leader than Nickens, who died in a car crash shortly before the 1984 negotiations.

In 1970, Wyandotte was bought out by the German chemical giant BASF (Badische Anilin und Soda-Fabrik), with the new company becoming known as BASF-Wyandotte Corporation (BWC). The BASF buyout of the Michigan firm reflected the increasing influence of multinational corporations in the American chemical industry. By 1986, for example, the globalization of the U.S. economy meant that 30 percent of OCAW members worked for foreign-owned companies. In December 1985, BASF dropped the Wyandotte name, although even in the 1990s some older workers still referred to the company as "Wyandotte."[23]

BASF has a long and turbulent history. The company had originally been set up in 1865 by Frederick Engelhorn, a jeweler, along the banks of the Rhine River at Mannheim, and it prospered in the late nineteenth century by being one of the first to manufacture dyes from coal tar. BASF pioneered a process where it took the tar, a messy by-product of gas distillation, and turned it into a product that replaced a more costly and unreliable organic substance. By the turn of the century, BASF had used the profits from selling these dyes to finance its diversification into inorganic chemicals and to construct new production facilities across the river at Ludwigshafen. In the early twentieth century, journalists called Ludwigshafen "The World's Greatest Chemical Works," as over 8,000 people were employed at the sprawling site.[24]

Around the turn of the century, BASF became part of a major chemical cartel. Two German chemical cartels, one based around BASF and Bayer, the other anchored by Hoechst, fixed prices, shared profits, and grew in power. Both cartels played an important role in supplying the German army during World War I, adapting their dye formulas to produce mustard gas and munitions. They also used their marketing power to limit the growth of the U.S. chemical industry. In 1925, the top executives in the chemical industry decided that further consolidation would cut inefficiency, eliminating duplicate product lines and separate sales forces. Hundreds of German chemical companies formally merged with BASF as a result. A new corporation, renamed Interessengemeinschaft Farbenindustrie, or I.G. Farben, was created. The BASF name consequently ceased to exist until 1952.[25]

I.G. Farben was an immensely powerful conglomerate that set quotas and pooled profits. Afraid that left-wing politicians might come to power and nationalize the cartel, I.G. Farben executives began to finance Adolf Hitler's National Socialist Party. The cartel profited by supporting the Nazi regime, and by 1942 it was making a yearly profit that was greater than 800 million marks more than its entire capitalization in 1925. During World War II, I.G. Farben was given possession of chemical companies in foreign lands such as Czechoslovakia, and these captured lands provided slave labor to the German factories. I.G. Farben plants were also built next to the concentration camps at Maidanek and Auschwitz. When the slave laborers could no longer work, they were moved to Birkenau and gassed by Zyklon B, the patent of which was owned by I.G. Farben. Due to the cartel's central role in the Nazi war machine, nearly half of the Ludwigshafen site was destroyed by Allied bombing in the course of World War II.[26]

After the war, many members of the I.G. Farben board were arrested and indicted for war crimes at the Nuremberg trials. After operating under Allied supervision from 1947 to 1952, I.G. Farben was divided into three large firms—Bayer, Hoechst, and BASF. As a separate company again, BASF set about rebuilding its damaged Ludwigshafen site. In the two decades following the ending of World War II, West Germany lacked money to import chemicals from abroad, and BASF consequently prospered by selling its products in the domestic market. The company grew rapidly and employed 45,000 workers by 1963, continuing to expand by marketing its line of magnetic audiotapes and then producing videotapes.[27]

With little room to grow further in Germany, the company began to cast an envious eye at the vast U.S. market. The takeover of Wyandotte meant that BASF now owned several U.S. chemical plants, including the large site in Geismar. Its appetite whetted, the German firm increasingly moved into the U.S. market over the next fifteen years. In 1978, it acquired the dyestuffs and intermediates plant and business of GAF Corporation in Rensselaer, New York, for $25 million, while three years later it bought Limbacher Paint and Color Works for an undisclosed sum. In 1978, BASF also purchased Dow's 50 percent holding in the chemical company Dow Badische. BASF's U.S. subsidiary established its headquarters in Parsippany, New Jersey, and by 1984 it was responsible for thirteen production sites and research facilities with annual sales of $1 billion. The Geismar works was, however, particularly important to the company.

3. The BASF complex from the air in the mid-1980s. (Courtesy PACE International Union)

"Geismar was, some folks like to say, the crown jewel of the corporation," noted Richard Donaldson, a former industrial relations manager at the plant. "It was the largest facility and certainly the most important," confirmed Bill Jenkins.[28]

The size of the site was particularly important to BASF executives, as they knew that they had plenty of opportunity for further expansion. Throughout the 1970s and early 1980s, in fact, BASF's operations in Geismar grew steadily. Soon after purchasing the Louisiana plant, BASF constructed a new plant to produce polyether polyols, and in 1974 it added a facility to make the herbicide Basagran. Six years later, the company constructed a plant to make MDI (methylene diphenyl isocyanate), which was used to produce rigid foam insulation. The site thrived, partly because most of the chemicals made at Geismar were specialty chemicals. The production of specialty chemicals was highly capital intensive and technology-based, ensuring that there was little competition from Third World countries. By 1984, the German chemical-maker was the largest employer in Ascension Parish.[29]

At the time, the change of ownership from Wyandotte to BASF mattered little to most workers. Although BASF was not a family firm, workers still felt a great deal of pride in working for the company. Roger

Arnold, who started at the plant in 1976, explained how many felt: "When you went to work, at that time it was known as BASF-Wyandotte, you'd go anyplace and you'd say, 'I work for BASF-Wyandotte.' You'd stick your chest out and you were proud, because that's the way it was, you had one of the top jobs in the country at that time."[30]

Workers felt this sense of pride largely because they were paid well. The chemical industry is highly capital intensive and rewards its workers accordingly. In 1975, for example, chemical industry workers in Louisiana earned nearly 40 percent above the general state factory average for comparable workers.[31] Workers at both Wyandotte and BASF were acutely aware of the fact that they earned much more than most other workers in the state at the time, and their jobs were therefore highly prized. Turnover was low. A seniority list compiled by BASF in February 1987, for example, showed that over 60 percent of workers had more than ten years of experience at the plant. The maintenance jobs, which were the highest-paying and most highly prized, were mostly held by senior workers in their forties and fifties.[32]

Workers' high wages also conditioned their attitude to environmental issues. Most reasoned, in fact, that the high wages they received were compensation for being exposed to dangerous chemicals. "We saw the wages as a trade-off for the danger," noted Carey Hawkins. "It was a trade-off, your health, the shift work, the unsteady hours, the extended hours. A lot of the times you wound up working sixteen-hour shifts and knowing that the chemicals that you're fooling with may have a future effect. . . . it was a trade-off for the wages. It was like the operations jobs, if you weren't fortunate enough to go onto college, financially or whatever, it was a good job. You would do well. For this area, it was probably the highest-paying jobs." Bobby Schneider recalled that he had a similar attitude: "I can remember telling my mother prior to this [the lockout], somebody had said something about hazardous waste, and I can remember making the comment, 'Well, if you want the good-paying jobs, you're just going to have to put up with it.'" Some workers accepted that working in a chemical plant could shorten their life expectancy. "We just knew that, as a general rule, chemical refinery workers don't live as long as other people but we thought, two or three years, what the devil, it was good money," explained John Daigle. "We figured in the long run it was worth it. You had a better life for a slightly shorter period."[33]

Daigle remembered that although workers were aware that the chemicals they were dealing with were dangerous, they did not think about the broader consequences of pollution, particularly its impact on the communities outside the plant: "We didn't think it got out of the plant. . . . It's not affecting anybody else. It's just us. And that was it. We didn't figure stuff that we dumped on the ground soaked into the ground, got into the water." Daigle added that workers "just believed what the company told us."[34]

Workers admitted that they often failed to challenge poor environmental practices, mainly through lack of knowledge. "I think it was ignorance, that people didn't realize the harm that they were doing," noted Hawkins. "Back then it was like the Mississippi River could wash away anything, but we didn't know and we didn't understand. . . . We were ignorant. We had these things that could cause tremendous harm that we were doing, these practices that were not environmentally safe, but we just did not know."[35]

Roy Fink, who started working at the plant in 1965, emphasized that workers accepted what the company had told them. "They just had the attitude, 'Look, don't question what I'm telling you to do, just go ahead and do it,'" he recalled. "It was stuff like dumping stuff to the ground or burning stuff at night. Burning residues and this type of stuff, you couldn't burn it during the day, you had to burn it after it got dark. Of course, we wasn't educated in none of that stuff. We just figured, well they knew what they were saying because this was a big company and we wasn't nothing but just country people. A supervisor would tell you, 'Oh, that ain't going to hurt nothing. Just go ahead and do it, don't question it.'" "We just never thought," added Esnard Gremillion. "It was just part of your job. . . . We was doing the work, like dumping the acid in the holes and all that kind of stuff and we didn't think there was anything wrong about it."[36]

Workers did not express support for the environmental movement that emerged in the 1960s and 1970s. In the early 1960s, Rachel Carson's *Silent Spring* had a major impact on public opinion, alerting Americans to the threat that pesticides and nuclear radiation posed to both wildlife and humans. Influenced by the political activism of the 1960s, grassroots environmental organizations expanded rapidly in the late 1960s and early 1970s. Several important groups were formed in the immediate aftermath of the first Earth Day, which occurred in April of 1970.[37] The emergence of these organizations was a reflection of mounting public concern about pollution; Gallup polls taken in 1965 and 1970, for example, recorded an increase from 17 percent to 53 percent in the number of people who rated "reducing pollution of air and water" as one of the three problems they would like the government to devote more attention to.[38]

As the environmental movement grew, it influenced the passage of a great deal of new legislation. In 1970, for example, Congress passed both the Clean Air Act and the Water Pollution Control Act. The former set ambient air quality goals that were supposed to be achieved by July 1, 1975 (although these were not met), while the latter established specific standards to limit the pollution content of wastewater effluents. The Environmental Pesticide Control Act of 1972, meanwhile, empowered the recently established Environmental Protection Agency (EPA) with the authority to regulate pesticide use. In 1970, the Occupational Safety and Health Administration (OSHA) was established, setting clear standards for exposure to harmful substances in the workplace. These standards were very difficult to enforce, however, especially in the chemical industry, where thousands of synthetic chemicals had not been tested for the effects of chronic exposure. Estimates of the portion of cancerous disease caused by exposure to chemicals among workers varied from 5 to 30 percent.[39]

The rise in environmental consciousness allowed older voluntary organizations such as the Sierra Club, which had been founded in 1892, to grow in size. At the same time, many new environmental groups emerged. When the Reagan administration relaxed environmental standards and slashed budgets for federal environmental agencies, many Americans responded by joining grassroots groups. Radical groups such as Greenpeace and Earth First! arose, pioneering direct-action tactics in order to protect wildlife and wilderness.[40]

BASF workers viewed the growing environmental movement with suspicion or even hostility. "We was more toward the company," explained Roger Arnold. "You know, 'That's just a bunch of environmental people that's trying to shut us down,' and stuff like that." Bobby Schneider recalled that he had viewed environmentalists as "a bunch of loonies that was out there. . . . they were after my job." Schneider added that prior to the lockout, most workers would have defended the company against environmentalists: "Up to that time, we probably would have went to bat for them."[41]

Among manufacturing workers, this outlook was typical. In the 1970s and 1980s, many workers opposed stronger environmental laws because they feared that they would cost them their jobs. Environmental groups, for their part, generally drew their membership from the urban middle class and had little contact with blue-collar workers. Some OCAW leaders claimed that environmentalists had shunned any collaboration with organized labor. "Some environmentalists," charged OCAW public relations director Ray Davidson, "have failed to recognize this need for partnership because some of them, to say it plainly, are intellectual snobs. From the heights of more extensive formal education, some of them look down on men who work with their hands and sometimes fracture English grammar."[42] Environmental groups accepted that most of their members were middle-class, but they denied that they were elitist. The mainstream environmental organizations, however, gave only passive support to the passage of OSHA, which was strongly supported by the OCAW's leadership, concentrating more on protecting wildlife and wilderness areas.[43]

Industry encouraged the schism. In the late 1970s and early 1980s, in particular, executives often blamed stricter environmental laws for plant shutdowns, insisting that they could no longer compete with products made in countries that lacked such laws. In 1980, for example, Atlantic Richfield Company shut down a copper plant in Great Falls, Montana, claiming that "the existing plant cannot be retrofitted to satisfy environmental standards and become cost competitive with modern large-scale smelters."[44] As a result, workers often responded to plant shutdowns by blaming environmental activists. "Environmentalists Are Polluting Our E-C-O-N-O-M-Y!" declared one bumper sticker popular with union members.[45] The OCAW itself was split over the issue, with some leaders supporting industry's position, while others argued that environmental laws did not cause plant shutdowns. Tony Mazzochi, the union's legislative director in the 1960s, repeatedly argued in favor of stronger environmental laws, but he recognized that not all the union's members supported him. "The industry is always able to wean the people away on the jobs issue," he admitted. When Mazzochi ran unsuccessfully for president of the union in 1979 and 1981, he claimed that his proenvironmental position lost him votes from members who were concerned that strict environmental laws would jeopardize their jobs. In reality, Mazzochi argued, environmental regulations were rarely the major cause of plant shutdowns, and he worked hard to communicate this to the union's members.[46]

Prior to the late 1970s, Local 4-620's members gave little thought to the environmental consequences of the chemical industry's growth in their state. As such, they reflected how most state residents felt at the time. As the petrochemical industry moved into southern Louisiana, local people did not think about the industry's ecological impact. Few laws existed to protect Louisiana's environment; there was, for instance, no legislation regulating the operation of landfills or hazardous waste sites, other than they were supposed to be operated in a "safe" manner. The state did not set up a Department of Environmental Quality (DEQ) until 1984. Prior to this, environmental quality was monitored by the Air Control Commission and the Stream Control Commission, both of which were largely ineffective. During the 1950s and 1960s, the Stream Control Commission regularly granted permits to companies that wanted to dump effluents in the Mississippi River. Willie Fontenot, head of the citizens' access unit in the Attorney General's Office, similarly remembered that the health department officials that ran the Air Control Commission often recorded air quality readings that were most favorable to industry, arguing that high levels of pollution were not typical. "The whole effort was to down play pollution problems," he charged. Although it was established in 1965, the Air Control Commission did not collect any penalties from industry until the DEQ was established. Even after this new state agency was set up, it was understaffed and underfunded. In 1985, Louisiana only had six inspectors to oversee environmental regulations at 38,700 oil and gas wells and 14,000 oilfield pits, many of them the site of severe environmental problems.[47]

In retrospective interviews, chemical industry executives admitted that insufficient thought was given to the environmental consequences of the industry's growth in Louisiana. Dan Borne, the president of the LCA since 1988, admitted that prior to the 1980s companies were "basically regulated only minimally." Richard Donaldson, who went to work in labor relations at BASF in 1972, felt that industry had needed to be more closely regulated. "We had a real large chemical waste problem in Louisiana because there was no real control over it," he asserted. Donaldson recalled that in his early days at BASF, managers were only too happy to wash their hands of hazardous waste. He pointed out that the favorable political climate in the state encouraged this attitude: "A lot of people said that Edwin Edwards was not very tough on industry, and he wasn't. He knew that the chemical industry was a backbone of the money in this part of the state, and Louisiana is a poor state, so he liked business coming in, he liked money coming in. So there was not real strong regulation then."[48]

Charles "Buddy" Roemer, a reformer who successfully ran against Edwards in 1987, was particularly critical of previous governors' environmental record. Widely regarded as the state's first "environmental" governor, Roemer felt that previous administrations had given a free hand to the petrochemical industry to pollute the state. "Jobs were the thing," he recalled. "We'll take your jobs no matter what kind of jobs they are. Come on down, man, we'll give you tax benefits, we'll subsidize you." Roemer felt that when he took office, Louisiana lacked any environmental standards. Roemer, of course, was a political opponent of Edwards, but his views are largely accurate. In 1990, Edwards himself admitted that the environment had been damaged as a result of the state's rapid industrialization in the decades immediately after World War II: "I am the only public official I know who has been honest enough to say, 'Yes, we made trade-offs.' We wanted jobs and we wanted industry and we allowed them to come in and operate based on technology as we knew it then. Now, we know better." Edwards defended his actions, however, claiming that they had provided much-needed "stimulation" to the state's economy.[49]

Further revelations about Edwards' links to the petrochemical interests emerged in 2000, when the former governor, along with seven others, including his former legislative aide and a state senator, went on trial on charges of extortion, racketeering, and fraud in connection with the granting of fourteen state gambling licenses. During the trial, it emerged that the Edwards' administration had issued a controversial permit to Marine Shale, a hazardous waste disposer located near Morgan City. Jack Kent, the owner of Marine Shale, made a $75,000 contribution to Edwards during the 1986–87 gubernatorial campaign.[50]

In many ways, Edwards is a reflection of what political historian Tony Badger has called the "persistent culture of corruption for public governance in Louisiana." Many prominent political figures in the state were suspected of corruption before Edwards, including former governors Huey and Earl Long. Like Edwards, for example, Earl Long probably received cash payments in return for allowing untrammeled slot machines in southern Louisiana. It became commonplace for political friends to be rewarded, and even into the 1990s stories about political corruption in the state frequently made the headlines. Between 1992 and 1995, for example, thirty New Orleans police officers were convicted and sent to prison, including the chief of the vice squad, who was convicted of robbing banks and strip clubs.[51]

As was the case across the United States, chemical plants were often located in Louisiana in areas where residents were poor and African-American. Several factors account for this, including cheap land values and the relative political powerlessness of poor communities. Prior to the 1980s, blacks tended to be less active in the environmental movement than whites, ensuring that corporations faced still less organized opposition when they located in these communities. Some studies, particularly those conducted by sociologists in the 1960s and 1970s, asserted that this was indicative of a lower level of environmental concern, although more recent work by environmental sociologists has also criticized the environmental movement that emerged in the 1960s for making insufficient effort to reach out to minorities.[52] In Louisiana, many residents initially welcomed the plants into their communities because they hoped that they would receive well-paying jobs in return. Albertha Hasten, an African-American environmental activist who lived immediately across the river from Geismar, remembered that black residents had supported industry at first. "When plants came in, we always thought about jobs, money," she recalled.[53]

The African-American residents who lived near the chemical plants soon found that their communities received few benefits from industry. The plants hired predominantly white workers from farther afield, overlooking African-Americans who lived just outside their gates. This hiring pattern started when industry first arrived. Like Wyandotte, many of the companies that located in Louisiana were northern-based, and they often imported the first workers from out of state, insisting that experienced staff were required to get the plants started up. Once the plants were running, they tended to follow the segregated hiring practices that prevailed in most southern industries until at least the mid-1960s, employing few African-American workers and restricting them to low-paying, labor-based jobs.[54]

The hiring practices of the BASF-Wyandotte plant typified this pattern. Until the early 1970s, the vast majority of workers at the plant were white and male. The small number of African-American workers were restricted to only two laboring jobs, known as "laborer" and "laborer pusher." Frank Smith, an African-American worker hired at the plant in 1967, recalled that when he started, "all blacks came through the labor gang." Blacks, noted another worker, performed all the "grunt work" in the plant.[55]

At Wyandotte, black applicants were required to pass preemployment tests that were continued by BASF after it took over the plant in 1970. Following complaints from African-American workers, the U.S. Justice Department investigated the plant in 1972 and judged the tests to be racially discriminatory. Indeed, on April 20, 1971, only 5.1 percent of workers at Geismar were black, even though Ascension Parish was around 30 percent black. The Justice Department investigation further noted that BASF was guilty of "racial discrimination with respect to assignment of new employees," adding that this practice had continued "until the past few months."[56]

In the company's defense, BASF manager Richard Donaldson stressed that segregated job assignments were common across the southern chemical industry at the time. Prior to the 1960s, indeed, black workers were confined to a limited number of laboring jobs in a wide variety of southern industries. Like many other companies, BASF made little effort to change its practices until pressure was exerted by both African-American workers and the federal government. Faced with a damning investigation from the Justice Department, the company entered into a consent decree to try and improve opportunities for black workers. The preemployment tests were abolished and the seniority system was changed to give African-American workers better promotional opportunities. Rather than being permanently consigned to laboring jobs, these workers were now allowed to use their plant seniority to transfer to better-paying positions. Like other managers at other companies, however, BASF managers worried that African-American workers would not be sufficiently qualified to perform the higher-paying jobs successfully, and they continued to defend their employment tests as a necessary screening device.[57]

After the consent decree, the job opportunities of African-American workers did improve, and they had steadily worked their way into production jobs by the early 1980s. At the time of the lockout, around 15 percent of the BASF workers in Geismar were African-American. This was, however, still well below the proportion of African-Americans in Ascension Parish, and many black residents complained that there was still hiring discrimination. Managers, however, insisted that they had to hire from farther afield in order to fill jobs in the plant, many of which required a high level of skill.[58]

As African-American workers began to be hired into production jobs, the Geismar plant also began to employ its first women workers. The number of women workers remained very low, however; at the time of the lockout, only 8 workers out of 372 were female.[59] In December 1978, Gladys Harvey was the first woman hired as a production worker, and she recalled that it took a great deal of determination to hold down her job: "It was awful. The men were very closed-minded in the beginning. They had an old man had been there about thirty-five years, and he just looked at me and said, 'Can't no damn woman do my job.' It was kind of tough, and then I got the wolf-whistles and all that good stuff that goes with being the only woman around." Harvey was determined to confront prejudice and prove that a woman could hold down a job at BASF. "I'm too stubborn to quit," she asserted. "I knew I could do the job and that it would take a long time to prove that and there would always be some of them that didn't want me there." She battled on, performing the physical "outside jobs" and eventually winning the grudging respect of male colleagues. The BASF plant remained male-dominated, however, and Harvey noted that many of the women hired after her had quit. "They just couldn't physically handle it," she noted. "They just couldn't do it. They decided it just wasn't worth it. It was a lot of strain, a lot of stress." A decade after the lockout, fewer than twenty women worked at BASF.[60]

Like most chemical workers in Louisiana, the workers at BASF's plant had little contact with the residents of Geismar. BASF continued Wyandotte's practice of hiring most of its workers from Baton Rouge, and by the early 1980s around 70 percent of the company's workers lived in that city rather than in Ascension Parish. The result was that most workers only drove through Geismar on their commute to and from work and had little contact with the community there. The union even met in a rented building in Baton Rouge rather than in Ascension Parish. As the union entered contract negotiations with BASF in 1984, therefore, it lacked any bond with the community.[61]

2

Negotiations

On June 1, 1984, representatives from BASF and the OCAW met in Baton Rouge to negotiate a new contract for the company's Geismar plant. The talks began with managers presenting the OCAW representatives with a raft of contract proposals that called for a wide range of concessions from the union, including a wage freeze, major changes in the seniority system, and an increase in workers' contribution to the health plan. "They just come out guns smoking," recalled Bobby Schneider. "It was obvious we was going to have a dispute or either have to take drastic cuts all the way." The union resisted the company's demands, and although the two sides continued to meet for the next two weeks, a dispute always looked likely.[1]

The company's conduct marked a clear change from the way that the two sides had previously conducted negotiations. Prior to 1984, the union had come to the bargaining table with a list of demands. The two sides had then bargained, with the company seeking to reduce the union's demands. A compromise would be reached that allowed both sides to feel satisfied; the union had secured some gains in wages and benefits to take back to its members, while the company had succeeded in limiting the cost of the union's claims. Richard Donaldson, the manager of human resources at Geismar since 1972, remembered that the company and union had a peaceful and constructive relationship until the 1984 negotiations: "There was no animosity between the company and the union before the lockout. I mean we took them really by surprise with this major overhaul of the contract. . . . Pretty much, labor negotiations have been the union comes in and says, 'Look guys, this is what we want to keep on working,' and you dicker with them until they agree that, 'Okay, we'll only take half of what we want, and we'll continue working,' and so this was a little harder stance. . . . So when we put our proposal on the table in 1984, we had some very, very radical stuff that they had never seen before, never dreamed that anybody would be awful enough to put that on the table, and they just didn't know what to do except to dig in and say, 'No.'"[2]

Before the lockout, the relationship between the two sides was, as Donaldson put it, a "traditional" labor-management relationship. In the 1960s and 1970s, labor relations across the United States largely followed the same pattern; unions made demands at contract negotiations, and the company then bargained to reduce the scale of these demands before reaching agreement. There was, some scholars have argued, an informal "compact" or "accord" between labor and management; management offered high wages and good benefits, and labor agreed not to strike in return. As was the case at BASF, however, this accord fell apart in the 1980s as companies became increasingly assertive at demanding concessions from unionized workers. Often citing increased foreign competition, many companies turned to their hourly employees for givebacks in wages, benefits, and well-established work rules.[3]

If unions responded by striking, many companies were willing to continue operating. After Ronald Reagan fired striking members of the Professional Air Traffic Controllers Organization (PATCO), companies increasingly responded to strikes by hiring permanent replacement workers, a tactic that labor historian Robert Zieger has termed "a virtual declaration of open warfare against the union."[4] Permanent replacements were a departure from the temporary substitutes that management had traditionally turned to during labor disputes before the 1980s. Unlike temporary replacements, who left at the end of a strike, permanent replacements were assured of strikers' jobs. After the strike, the law gave strikers the right to return to their old jobs, but only if replacements left them.[5] As workers became afraid of losing their jobs, the strike rate plummeted throughout the 1980s. Between 1978 and 1992, for example, the number of strikes involving more than 1,000 workers fell from 235 to 35.[6]

In the early 1980s, BASF, like many companies in the United States, decided to take an increasingly hard line with its unions. Although they worked for a German-owned company, U.S. managers recalled that they were given complete autonomy in devising the company's labor relations strategies, with no interference from Ludwigshafen. Les Story, who became the site manager at Geismar in 1979, recalled that goals for negotiations were set by corporate management in Parsippany, New Jersey, in consultation with local site management. "The human resource policies of the American company were determined in America," he asserted, "and the policy of Ludwigshafen had always been that these were local issues.

. . . I never got any instructions from Germany." "The U.S. company ran very autonomously," added Richard Donaldson.[7]

The company's 1984 proposal reflected the increasingly hard line that a number of key U.S. managers were taking in dealing with unions. Since 1979, the president of BASF in North America had been Edwin Stenzel, who had moved to the German chemical maker from Dow Chemical Company. Dow was a largely nonunion company and had a reputation in labor circles as antiunion. Some BASF managers shared this opinion. "Dow had a history of being anti-union from my perception," noted Bill Jenkins. Jenkins added that when Dow had built new facilities while Stenzel had been president "they were very careful to build them salaried, to keep them nonunion." Stenzel himself refused to comment publicly about his views toward unions, but Jenkins noted that the former Dow executive had introduced a new "toughness" into the company's dealings with organized labor.[8]

This toughness was expressed in the bulky contract proposal that corporate management drew up for negotiations at Geismar. The author of the proposal was Henry Kramer, BASF's corporate manager of labor relations. Kramer was committed to reducing union influence in BASF's plants, and other company executives admitted that he strongly disliked unions. "I like Henry a lot," reflected Bill Jenkins. "Henry was a wild man, though. Henry did not like unions. Henry disliked unions very much, philosophically. So for Henry, to be real honest with you, Henry became kind of a tool—he was kind of like a gunslinger. If we wanted to scare the union or cause havoc, we'd send Henry in." Richard Donaldson described Kramer as "a mean, mean little dog" who was the "main pusher" behind a drive to reduce union influence in BASF plants. A short, quick-tempered man, Kramer was a trained lawyer who was given the main responsibility for drawing up contract proposals. "You see a contract for a lease, for instance," noted Donaldson. "That's written for the person that owns the property. It's not written for the person that rents the property, and that's the way Henry would write contract language. He was not a benevolent person."[9]

Kramer himself, in retrospective correspondence, denied that he was "antiunion," asserting that his views were "far more complex" than this term implied. He insisted that unions were "largely unnecessary," however, because progressive companies could provide the same benefits without them. He also claimed that unions were "internally undemocratic" organizations that used "violence and intimidation to keep their members in line." He admitted that the labor relations policy he pushed at BASF was "not union oriented." Believing that the company could pro- vide the same benefits as unions, Kramer's ultimate aim was to create an excellent relationship with BASF employees. They would then voluntarily vote to decertify the union, especially if they saw that they could receive better benefits as salaried workers.[10]

In April 1983, BASF's labor relations managers met in Williamsburg, Virginia, for a labor relations strategy session that drew on these ideas. The company admitted that its long-term goal was to be "union-free" and argued that this could be achieved by preventing unions from gaining further concessions from management. As a company presentation put it: "This long-term objective of being union free means we are determined to make unions unnecessary—and this causes us to deal with unions in a determined way." The presentation, apparently given by Kramer, put forward the so-called Jelly-Bean Theory. This compared labor-management relations to an encounter between a bear and a car of visitors in Yellowstone National Park. When the bear (the union) approached the car (the company), the visitors gave the bear a jelly bean, encouraging the animal to look for more. When the visitors' jar was empty, the bear became aggressive, expecting that assertive behavior would continue to be rewarded. The presentation argued that companies who gave concessions too easily to unions were thus like the visitors who gave all their jelly beans to the bear. A better approach was to take a hard line in negotiations, thus destroying the idea that unions could deliver benefits to their members. Union members would become disillusioned with the inability of their unions to provide economic benefits. The company could then stress to their employees the benefits of becoming salaried, as this would offer them better benefits, and the local union would cease to exist.[11]

This goal of becoming salaried was a central part of BASF's strategy. The chemical industry has always required a large number of salaried workers, and at Geismar around half of the eight-hundred-person workforce was already employed on this basis in the early 1980s. BASF aimed to persuade union-represented employees to also become salaried, thus operating the entire plant on a nonunion footing. A memorandum sent to Les Story from the company's U.S. headquarters in June 1983, for example, called for the Geismar manager to "make our goal of an all salaried operation known to all." Managers were encouraged to explain the "advantages" of a salaried operation to union workers, particularly the superior benefit package available.[12] BASF's new facilities were all run on a salaried basis. In the early 1980s, workers at several BASF plants, including the company's large facility in Wyandotte, Michigan, decertified their local unions and became salaried.[13]

OCAW leaders saw these decertifications as evidence of a coordinated union-busting strategy, while BASF executives were keen to stress that the workers exercised their own choice to be union-free. Bill Jenkins asserted that in Wyandotte, for example, it was union officials who initiated the decertification, and they did so because they were attracted by the benefits of being salaried: "At Wyandotte, there was an element in the union, their officers, that had come to the conclusion that they wanted better benefits, and what they wanted were the salaried benefits, and they had brothers and sisters that worked for the company that had better benefits, and they came to the conclusion the way to do that was to decertify the union." Henry Kramer similarly argued that "in those locations where there had been decertifications, the unions had not 'folded,' they were simply run out by the employees under a legal system in which the company cannot and did not have any part. Their unions gave up because they had insufficient support for being retained." At other BASF plants, local unions continued to exist but only by accepting concessionary contracts. In Huntington, West Virginia, for example, an OCAW local was locked out by BASF for 281 days in 1981 before the two sides eventually secured a new contract on the company's terms.[14]

As BASF's managers approached the Geismar negotiations, they brought with them a new determination to change the nature of traditional bargaining and exact concessions from the union. Executives insisted that they had operated on the "Jelly-Bean Model" for too long, giving in too easily to union demands: "Over many years, Wyandotte had a history of rewarding this kind of behavior and giving into union demands without ever having any demands of their own."[15] In the summer of 1983, the company had signaled its new approach when, on the advice of Kramer, it had told Local 4-620's chairperson that the union could no longer conduct its business on company time. Since 1978, the chairperson had been allowed to carry out union affairs for twenty hours a week and had been provided an office by the company. In July 1983, however, Esnard Gremillion was informed by BASF managers that the office was no longer available to him. The union filed a complaint with the National Labor Relations Board (NLRB), which ruled that the company had been wrong to make this change "unilaterally" and should reinstate the office, but by this time the lockout had already begun.[16]

Prior to negotiations, corporate managers met with local management and asked them what kind of concessions they would like to secure from the union. Work rules quickly emerged as an area where both corporate and line managers wanted to see changes. After listening to line managers, the company proposed a "complete replacement" of the seniority article.[17] Both corporate executives and local management felt that the existing seniority provisions were too permissive because they allowed workers to bump on the basis of seniority without necessarily being qualified to perform the jobs. The company felt strongly about securing these changes and was, according to Bill Jenkins, prepared to endure a strike in order to achieve them.[18]

The company also sought to roll back wages. In 1981, BASF had agreed to give union members substantial wage increases at a time of high inflation. By 1984, however, inflation had fallen and the company argued that workers were now too highly paid. In 1980, national inflation stood at 12 percent, but it dropped to 4 percent by 1983 and held steady thereafter.[19] The major competitors for the products that BASF made in Geismar were U.S. chemical producers, and managers were keen to ensure that they did not pay more than their competition. Managers were also very aware that most of their competitors were nonunion and that these firms had been able to respond to the fall in inflation by unilaterally adjusting their wages on a yearly basis. BASF, in contrast, was committed to a wage increase over a three-year agreement and felt disadvantaged as a result. Data collected by executives showed that the German chemical-maker paid its first-class operators in Geismar $14.80 an hour, thirty-four cents an hour more than nonunion plants in the area. Although BASF workers usually had been paid a little less than their nonunion counterparts in the past, the fall in inflation meant that their wages were now slightly higher.[20]

In 1983, the company approached the union and asked it if its members would reduce or forgo their third-year wage increase because of the fall in inflation. Managers had already imposed a wage freeze on salaried employees and argued that hourly workers should follow suit. Ernie Rousselle met with BASF executives and inquired about what the company intended to do with the money that it saved from not having the wage increase, suggesting that at least part of it should be set aside to compensate workers in the event of a layoff. BASF, however, refused to consider the proposal. A piqued Rousselle responded by telling the company that the union would not agree to surrender the wage increase. BASF therefore came into the 1984 negotiations seeking an eight-month wage freeze.[21]

In the early 1980s, many companies, especially nonunion firms, were also seeking to increase the percentage of health care costs that was paid by workers. BASF surveyed most of the companies along the chemical corridor and concluded that it needed to secure increased employee contributions to the company's health care plan in order to remain competitive. The company's health care proposal called for an 80/20 coinsurance plan with an out-of-pocket limit on hourly employee expenses. Executives had already persuaded their salaried employees to agree to paying a share of their health care costs, and they felt that union workers would have to accept the same change. If they did not, the company's strategy to reward salaried workers more generously than hourly workers would be undermined. "We had already implemented for the salaried workforce some cost-sharing and some changes in benefits, and we were basically asking the union to go along with the same thing that everybody else in the company had gone along with," explained Bill Jenkins. "And that was a strike issue for us; we couldn't have any facility that had better benefits than all the salaried workers and the other unionized workers in the company. That would lead to an unfair situation and a big problem."[22]

The union, however, rejected all of the company's proposals. On seniority, the differences between the two sides were marked; the union wanted to maintain contract provisions that allowed senior workers in a unit faced with layoffs or shutdown to bump junior workers in another, rejecting BASF's contention that the retraining necessitated by bumping was excessively costly and reduced worker productivity. Local 4-620's members were prepared to stand firm and fight to uphold the contract's seniority provisions because they felt, like many union members, that the provision of seniority was a central purpose of any union. They valued their union primarily because they felt that it made the workplace fairer, eliminating discrimination and favoritism. By seeking a major change in the seniority system, union members felt that BASF was striking a blow at the very heart of the union. "The first thing they done was try to take away your seniority," noted Roger Arnold. "Well if they take away your seniority, what they do, they break the union." Seniority rights were, as Marion "Putsy" Braud put it, "sacred things to a union person." For many workers, maintaining the contract's seniority provisions became the primary issue in the dispute. "The main issue is *seniority*," noted one union document, adding that workers were "united 100% to preserve the seniority system."[23]

The union rejected the company's wage freeze, initially proposing an 8 percent increase instead, and they refused to discuss the proposed health care plan changes.[24] Serious disagreement between the two sides was apparent from the very start of the negotiations. On the first day of talks, for example, Ernie Rousselle told the company that their proposal was completely unacceptable to the union: "We feel that what you have on the table completely disrupts what we have negotiated over a period of years with the Company."[25] A traditional labor negotiator who was proud of securing high wages and good benefits for his members, Rousselle was not prepared to accept wholesale concessions. The company, however, repeatedly insisted that its changes were necessary in order to make the plant more competitive and ensure its long-term viability. Executives claimed that they were on a "mission" to "get control of our costs" in order to ensure the "future of the site." Although the two sides met more than fifty times, they made very little progress, with neither adversary willing to move from the positions marked out at the start of talks, particularly regarding seniority.[26]

Poor personal chemistry contributed to the failure to reach a settlement. The 1984 negotiations were the first occasion where Ernie Rousselle and Les Story had negotiated face-to-face. A native of Chicago who had previously worked for BASF in Michigan, Story had only arrived in Geismar in 1979, and the 1981 negotiations had been handled by another OCAW representative while Rousselle served as a vice-president of the union. The two men failed to hit it off, with both regarding the other as too inflexible. They quickly took a personal dislike to one another, and even in retrospective accounts had a tendency to blame each other for the dispute. In previous negotiations, the two sides had been able to make progress away from the table, often through social meetings. Despite clashing over the table, for example, Jenkins and Rousselle respected each other on a personal basis, and Jenkins had even gone out for a beer with Rousselle to celebrate the settlement of the 1978 contract. In 1978, BASF had also locked out Local 4-620's members after the company had refused to agree to all of the union's economic demands. After less than a week, however, private meetings between Rousselle and Jenkins led to a breakthrough, and the brief dispute left little lasting bitterness.[27]

Story, in contrast, never established any communication with the union away from the bargaining table, and he later accepted that this had helped to lead to the dispute. "I regret my own personal inability to develop what we called a side-bar communication with the union leaders during the negotiation," he recalled. "In the 1981 negotiations, both Bill [Jenkins] and I and Richard [Donaldson] had all been able to have private communications where we would check the status of things that were going on and it was a much, much less cordial working atmosphere. It should never be a friendly atmosphere when you get down to negotiations but it got out of control, and I lay a lot of the blame for that at Rousselle's feet. . . . you know yourself in government how much secret negotiation on the side sometimes provides the solution to the problem, private conversations." Story nevertheless accepted his own share of the blame for this breakdown in communication. "That has to be fifty percent my fault," he acknowledged.[28]

Despite disagreements over specific issues, union members felt that the real issue in the dispute was the company's desire to break the union. "They wanted to do away with the union, and they . . . would run the whole show," asserted Esnard Gremillion. All the workers that went through the dispute insisted that the company wanted to run the workplace free from union interference. "They didn't want the union," recalled Roger Arnold. "They want flexibility to do what they want to, when they want to, how they want to. Not to worry about anybody or anything."[29]

Many outside observers shared this view. Political leaders who encountered BASF executives came away with the impression that the company was trying to break the union. Buddy Roemer, who met with company executives while he was governor, was a conservative Democrat who later became a moderate Republican. Roemer was not a strong union supporter, but he had little doubt about the company's goals: "To beat the union was their aim, to break the union was their aim, to bust the union, I think they put it to me, was their aim, and in America, you're allowed to have those aims, but you're not guaranteed to get them." Louisiana attorney general William Guste similarly felt that the lockout "took place for one basic reason, BASF wanted to break the union. It wanted to get rid of the union in its plants, and so it locked them out for that purpose."[30]

Both OCAW leaders and union members asserted, in particular, that BASF was taking advantage of the conservative political climate. "The company took a position across the bargaining table, 'We've got a Republican in the White House who's antiunion and we're going to kick your ass,'" charged Ernie Rousselle. Executives, in contrast, generally tried to downplay the influence of the political climate, insisting that the concessions they sought were vital in order to improve the economic competitiveness of the site. "Our overall objective was to form a basis for keeping the Geismar site competitive and productive," asserted Henry Kramer.[31]

Like Kramer, BASF managers refuted the assertion that they were trying to break the union at Geismar by arguing that their only concern was the economic competitiveness of the site. In reality, however, these two issues were inextricably linked, as managers felt that the site would be more competitive if union influence was reduced or eliminated altogether. According to Bill Jenkins, nonunion, salaried plants were much more efficient: "We had some salaried operations that were nonunion. They were a lot more efficient, they were a lot better run. They were better trained. We had almost no turnover in those kind of facilities. In the chemical industry, which relies upon skilled people, even at the operating unit level, it just makes a lot more sense to have salaried people that are part of the team." Les Story shared these views, asserting that he had always felt that the plant would operate better without a union because it would allow for more flexibility in assigning work. Story also explained that in the early 1980s, one of his major concerns was that BASF's non-union plant in Freeport, Texas, was receiving more investment from the parent company than Geismar. Thus, increasing economic competitiveness and reducing union influence were inextricably linked.[32]

At the Williamsburg strategy session, BASF managers also argued that union influence needed to be eliminated in order to increase the competitiveness of the company. The company stressed that the "costs" of having unions included the increased possibility of strikes and slowdowns, as well as the time that had to be spent on complaints, grievances, and contract negotiations. Overall, the company asserted that "life without unions is more productive, more efficient, and gives us more of an edge on our competition in the marketplace."[33] U.S. managers supported this position by circulating literature amongst themselves that argued that it was more profitable to be "union-free."[34]

Although BASF managers tried to downplay the influence of the conservative political climate on their actions, some evidence suggests that they were affected by it. In negotiations during the lockout, for example, management told the union that the company now possessed greater "leverage" to exact concessions, an apparent reference to the more favorable political climate for corporations. Les Story made a reference to the 1971 strike in one letter to the local union: "Although the union had the leverage in 1971, the Company has it now and is committed to solving problems caused by an unproductive labor agreement."[35] Ernie Rousselle also claimed that Bill Jenkins had boasted during the negotiations that the company now had "the stick" and planned to "beat the shit" out of the union. Rousselle argued that BASF was keen to exploit a political and economic climate that was favorable to union-busting. Jenkins remembered making these comments and asserted that he was trying to convince Rousselle that the company was determined to secure its bargaining goals and would not be intimidated. "What I said to Ernie was, I said, 'Ernie, you don't have the leverage this time. If you think that taking the guys out on strike or they're being withheld from working here is going to make any difference, it isn't going to make any difference. You have no leverage. We're prepared to run the facility without you.'"[36]

Part of the "leverage" that executives felt they had, as Jenkins pointed out, was the ability to operate the plant. BASF managers made preparations for a dispute well in advance of the lockout. In March 1984, for example, they circulated an emergency work pay policy to all salaried workers. If a dispute occurred, these workers were expected to take over production jobs and work long shifts, receiving overtime bonuses in return.[37] In addition, the lockout occurred at a time when unemployment in Louisiana was high, as the petrochemical industry was buffeted by both the nationwide recession and rising energy costs. During 1983, more than 3,000 chemical industry jobs were lost in the state. BASF executives therefore knew that it would not be hard to recruit replacement workers to run their plant.[38]

Management in the chemical industry as a whole was, moreover, clearly affected by the change in political climate that the Reagan administration represented. The pages of industry journal *Chemical Week* expressed unconditional support for the new Republican president. In January 1982, for example, an article written by one CEO praised the Reagan administration and its "positive steps . . . to remove federal interference with market forces."[39] It is also clear that industry officials were increasingly aware of the declining strength of unions, particularly the OCAW. Six months before the lockout in Geismar, *Chemical Week* noted that concessionary bargaining was becoming increasingly common in the chemical industry. The journal closely documented the OCAW's membership decline: "The union's membership has dropped 18% in three years to 130,000. OCAW's power has waned. . . . companies have repeatedly demonstrated their ability to run plants through strikes."[40] Throughout the early 1980s, in fact, the industry journal repeatedly commented on the weakness of the "troubled" OCAW, noting that the union had struggled to overcome the effects of the recession, which had made its members afraid of being militant.[41] It was not just the OCAW that had become "more passive"; *Chemical Week* also noted that even the Teamsters, which had obtained "years of double-digit annual raises and hefty hikes in benefits," were now yielding concessions.[42]

To a certain extent, *Chemical Week* was merely reporting what was happening across the industry, but the way that it highlighted declining union strength may have also influenced other companies to seek concessions. The journal was widely read by industry executives, and Les Story acknowledged that BASF managers were influenced by the concessionary trend: "We did look at a lot of situations around the United States that were occurring at the time, where wages were a big issue, where health care containment was a big issue, and we saw companies like ourselves taking a tougher and tougher stance against that. So in that respect we did look at that and we used that in some of the comparisons we showed to the union. 'Hey, this company did it, why don't you do it?'"[43]

In the decade prior to the lockout, business interests in Louisiana had also been very successful at restricting the influence of organized labor. In 1975 business groups had banded together to form the Louisiana Association of Business and Industry (LABI), which provided them with a more unified voice than ever before. Created out of the merger of the Louisiana Manufacturers Association, the Louisiana State Chamber of Commerce, and the Louisiana Political Education Council, the LABI was headed by Ed Steimel, an accomplished lobbyist with a public relations background. After arriving in Louisiana in 1979, Les Story also played a leading role in the new organization, serving as chairman of the Board of Directors.

In the fall of 1975, the LABI started a campaign to pass a right-to-work law, pointing out that Louisiana was the only southern state that had yet to pass such legislation. Enacted under Section 14 (b) of the Taft-Hartley Act of 1947, these laws authorized states to ban the closed shop (or compulsory unionism). Led by Steimel, the LABI carried out a sophisticated campaign to push through the law. At the heart of its campaign were dramatic thirty-second advertising spots on television stations that helped to swing public opinion behind the bill. Organized labor was, as the LABI put it, "caught off balance" by these tactics and was unable to respond effectively. As the LABI noted, the passage of the right-to-work bill signaled a new balance of power in Louisiana politics: "A new day seemed to be dawning in Louisiana, as business, professional and industrial leaders began to learn the value of unity and involvement in politics—a lesson unions had learned decades ago."[44]

To further confirm this fact, seven other pieces of labor legislation, most of them restricting union rights, were passed during the 1976 session. They included reform of the state's workmen's compensation laws and the repeal of the prevailing wage law, which had required government construction projects to pay the wage levels of union contractors. Organized labor was, moreover, unable to repeal any of the new laws. In the 1950s, Louisiana's unions had led a successful effort to overturn a right-to-work law, but they were unable to repeat their success two decades later, despite repeated efforts.[45]

Both Local 4-620's members and the OCAW's leadership were acutely aware of the increasingly assertive posture being taken by business across the country. Despite being unable to reach an agreement in 1984, the local union never considered striking BASF. Both union leaders and workers knew that an increasing number of companies were responding to strikes by hiring permanent replacement workers. In 1983, permanent replacements had been hired by the Phelps Dodge Company in Arizona, and the United Steelworkers, one of America's largest unions, went down to a crushing defeat.[46] If workers refused to strike, they knew that they could not be permanently replaced, as under U.S. labor law companies could only hire temporary replacements if workers were locked out. Newly elected OCAW president Joseph Misbrener dismissed striking BASF as "unthinkable"; it was, he said, "like lending a match to an arsonist."[47] Like Local 4-620's members, OCAW leaders were aware of the high unemployment rate in Louisiana and knew that if they struck, the company would have been able to recruit unemployed or lower-paid workers to take permanent jobs in the plant.

With the union unwilling to strike, a lockout became likely. On June 13, two days before the expiration of the contract, Bill Jenkins told the union that its members would be locked out if no contract was signed. Jenkins claimed that this course of action was necessary to prevent acts of sabotage by disgruntled workers: "We have a chemical facility, and we know we have a number of dangerous chemicals there and an employee that is upset could cause a severe horror." Jenkins worried that a few "hotheads" would carry out sabotage because they felt that the company had not acted fairly. He claimed that some incidents had already occurred, although the company never produced any compelling evidence of sabotage.[48]

Les Story stressed that the lockout was consistent with BASF's policy of not allowing unionized workers to work in the plant without a contract. Story asserted that this policy was "well-known in advance." The company had indeed locked out workers in 1978 on the same grounds.[49] Story was also concerned that if union members carried on working, the dispute would interfere with the safe running of the plant. He cited the experience of the Exxon plant in Baton Rouge, which was operating without a union contract at the time, to prove his point. "There is another concern," he noted in negotiations. "If our people have their mind on the contract not being settled, they will not keep their mind on the plant and what they are doing. Exxon has had a number of OSHA recordable injuries because their people are working without a contract. We do not want that to happen at Geismar."[50]

The company's concern about sabotage also ensured that they implemented the lockout a few hours before the expiration of the contract. Afraid that union workers would damage the plant just before leaving, managers caught workers by surprise and escorted them out of the gates several hours before the deadline. This move was resented by the union, which felt that negotiations should have continued right up to the five P.M. deadline. There was considerable bitterness as the two sides parted, with Rousselle claiming that "This committee has done everything they could to get a settlement despite the fact that we have had 12 months of the worst anti-unionanamous [sic] activities I have ever seen by the company." A long dispute already looked likely.[51]

3

They've Got Us Cornered

The first year of the dispute was a difficult one for the union. The OCAW did not launch its corporate campaign until the fall of 1985, and the local union struggled to exert leverage against BASF on its own. "They were kicking our butt," reflected Richard Miller, the OCAW strategist who was to play a key role in the eventual corporate campaign. The dispute failed to secure much press coverage, even in Louisiana, and locked-out workers struggled to secure well-paying work. The local union also made a strategic mistake by attempting to use the company's war record against it; this alienated the German unions and helped to destroy the international union's early efforts to bring pressure to bear on BASF.[1]

Shortly after BASF had escorted OCAW members out of the plant, Ernie Rousselle filed unfair labor practice charges against the company, asserting that they had not bargained in good faith throughout the negotiations. For Rousselle, the company's implementation of the lockout before the five P.M. deadline was indicative of their bad faith. Many workers initially placed their hopes in these charges and reasoned that they would soon be allowed to return to their jobs. In July 1984, however, the NLRB accepted the company's justification for the lockout and ruled that it was legal. Citing the many meetings between the two sides, the labor board asserted that BASF had bargained in good faith. For the union, the decision reflected the increasingly antiunion position taken by the board in the 1980s, a product of President Reagan's conservative appointments to that federal body. To BASF managers, however, the decision was a validation, and it encouraged them to continue taking a firm line against the union.[2]

For the first few weeks of the lockout, many workers used their free time to go hunting and fishing. Once this honeymoon period was over, however, the immediate priority was to find a job. With Louisiana suffering from a high unemployment rate, the vast majority of Local 4-620's members struggled to find work that paid well. In addition, many employers feared that the dispute could end at any time, and they did not want to risk hiring a locked-out worker on a full-time basis. As a result, most union members found themselves working two or three part-time jobs in an effort to make ends meet. Roy Fink recalled that he "wound up working three jobs" during the lockout, performing a variety of welding and pipe-fitting work. Frank Smith, in contrast, did secure a full-time day job, but it was low-paying, and he supplemented his income by cutting grass in his spare time. Other workers learned new skills in order to find work. Carey Hawkins, for example, trained as a swimming instructor and taught preschool classes at the YMCA, as well as working as a janitor in the evening and performing occasional construction work. The lockout, he recalled, was a constant struggle to "keep going." In order to find work, some workers ended up traveling long distances or left Louisiana altogether. Several Local 4-620 members, for example, found work near New Orleans and lodged together during the week in order to cut costs.[3]

Even with two or three jobs, however, most workers found that they were unable to make as much money as they had when they had worked at BASF. They had to make many sacrifices in order to survive, and they looked back on the lockout with sadness, relating that their families missed out on many experiences as a result. Roger Arnold, a heavily built maintenance worker, became emotional when recalling the dispute. Unsuccessful in his search for a full-time job, Arnold recalled that he was unable to provide for his children as well as he would have liked:

> My kids had it bad because they was used to a certain lifestyle and then that lifestyle was jerked out underneath them, and they couldn't have that lifestyle no more. Like my daughter, when she graduated from junior high, they had their little trip that they was going to take to Boston. Well, I didn't have the money to send her, but she got out there and she done work, cutting grass and everything like that, to make that money to go to Boston, and so she was able to make that by the people helping her out. . . . Most of their life was living this lockout, you know. This is all they could see. They couldn't get the clothes that the other kids was getting. They couldn't get the shoes that the other kids was getting, that was in their school. What I did to help out is coach the kids at school for volleyball or basketball, football. I tried to get involved with all the kids that I could at that time to help out, and kind of keep me occupied, to keep me from losing my mind, fighting this lockout all the time.[4]

Like Arnold, many workers fell into debt. By July 1985, the union calculated that locked-out workers owed $156,500 because they had been unable to meet their financial commitments. Twenty cars had been repossessed and thirty-seven families had either lost their homes or been forced to move to other housing.[5]

Despite these problems, Local 4-620's members did their best to hold together. The OCAW paid workers fifty dollars a week, and this helped many to stay afloat. In addition, locked-out workers received state unemployment pay for the first twenty-six weeks of the dispute. After lobbying from Ernie Rousselle, workers secured an additional thirteen weeks of this benefit because of Louisiana's high unemployment rate at the time. The union distributed its benefits every Monday night, and Rousselle recalled that the meetings provided a way of sustaining morale and reassuring the membership of the international union's commitment: "We paid lockout benefits every Monday, so we had constant contact with all of our people, and we had the whole committee there for the pay-out, so that a person could talk to his or her representative and come there every Monday," he recalled.[6]

Many workers also took comfort in the way that the dispute brought them together, relating that locked-out workers went to great lengths to find work for each other and to help each other out. "The guys that were locked out," reflected one worker, "in any way you could, you would help, help your friend get a job. Try to put a dollar in his pocket. . . . There was a certain pride that we all were sharing it, you know, and helping each other. 'Man, I heard there's a job over here,' or 'I have a job, you can come.'" Workers also felt that they were fighting for the life of their union, and they were not prepared to give in easily. "We were fighting for seniority rights and a lot of different big issues that mean a lot to union people," recalled Tommy Landaiche. As would be the case throughout the dispute, the company's demands for wholesale concessions, and its willingness to lock Local 4-620's members out of their jobs in order to try and secure them, united the workers and made them determined to resist.[7]

A particularly important move was the organization of the Women's Support Group (WSG). In the early stages of the dispute, many of the locked-out workers' wives felt isolated and uninformed about the lockout. The weekly union meetings had traditionally been attended only by union members, and many wives did not feel welcome there. Unaware of the issues of the dispute, some wives began to pressure their husbands to try and return to work. In August 1984, locked-out worker Paul Wiltz broke

4. A Local 4-620 union meeting, early in the lockout. (Courtesy PACE International Union)

with tradition by inviting his wife Rebecca to attend a union meeting. By going to the meeting, Rebecca Wiltz became informed of the issues of the dispute and, with Rousselle's encouragement, decided to form the WSG so that wives could both share their problems and work in support of their husbands.

The organization of the support group was vital because it mobilized the workers' wives and families behind the union. "I don't think the group could have held together without the support group, because when the lockout first started, the wives didn't have any way of knowing what the lockout entailed," recalled Rebecca Wiltz. "The women weren't allowed to go to the union meeting or they didn't go to the union meeting, and most of the husbands didn't keep the wives informed about what was going on. I think like my husband told me whenever it happened that it was going to last for three or four days and then they would be back in the plant. I think if the (Women's Support) group wouldn't have formed, then we would have encouraged the husbands to go back and cross the picket line and go back into work, because we wanted our paychecks for our families."[8]

A core of women became strong leaders in the WSG and proved very effective in communicating the union's case to a broader audience. In

March 1985, for example, WSG member Connie Kearns addressed the Louisiana AFL-CIO's convention to tell delegates about the struggle of the local union in Geismar. Her address highlighted the way that women were able to effectively use their positions as wives and mothers to communicate the economic impact of the lockout: "During this lockout, which has lasted over 9 months, many of these 370 families have lost their automobiles, their homes, and their life savings. As wives, it is heart breaking to see our husbands so degraded. As mothers, we have shed many tears as we see our children without proper food, clothing, and medical care."[9] WSG members also took the lead in writing letters to politicians and community leaders in an effort to build support for the union's case.[10]

In communicating the locked-out workers' cause to a broader audience, WSG leaders frequently clothed their demands in the language of Americanism. Playing on the company's foreign origins and its "dubious past reputation," they claimed, in one widely distributed WSG document, that the union was fighting to uphold "the American way of life." They portrayed the workers as citizens fighting tyranny, quoting William Jennings Bryan's dictum that "Clad in the armor of a righteous cause, the humblest citizen takes on a surprising strength." "We want our community to know, and we want BASF to know, that *not one* precious American freedom will be lost here," they asserted.[11]

As in the Flint, Michigan, sitdown strikes of 1937, where women marched outside the shut-down GM plants, the Geismar women also protested outside the plant on a regular basis.[12] Until the organization of the WSG, local union members had carried out few protests outside the plant. Union leaders were keen for the public to understand that Local 4-620 was not on strike but had been locked out, and they worried that picketing the gates would mislead a public and media that still often falsely reported that the Geismar workers were on strike. Once the WSG organized, however, its members began to regularly picket outside the plant gates. BASF responded by offering the women cold drinks, but even on hot days, when they were desperate for liquid, the determined protesters refused to take them. Through such public displays of support, WSG leaders felt that they had broadened the basis of the dispute beyond the locked-out workers. "One of the things the women brought was it showed the corporations we were behind our husbands and let the nation know too that it's not just a bunch of men out there," commented Wiltz.[13]

Under Wiltz's leadership, the WSG quickly grew in strength, and its meetings were regularly attended by over one hundred women. Wiltz's

5. Members of the Women's Support Group protesting outside the BASF plant, June 1985. (Courtesy PACE International Union)

activism took its toll on her personal life, however. Looking back on the dispute, she regretted that she had not spent more time with her family: "I was asking my daughter last night, that's thirty now, I said, 'Well what's the hardest part of the lockout for you?' and she said, 'The thing that Mamma and Daddy wasn't there for us.' Paul worked out of town and had to live out of town. I was highly involved in the support group, and they saw us worrying and everything, and when I was home, a lot of time I was making phone calls. It was just a different kind of lifestyle than I had ever been exposed to." Close to exhaustion, Wiltz eventually stepped down from the leadership of the WSG in 1987.[14]

Recognizing the valuable contribution that the WSG was making, most locked-out workers supported the group and were pleased when their wives participated in it. There were a few men, however, who wanted to keep women out of the dispute. "Some of them, their husband told them 'no' and they listened," recalled Laura Nordstrom, another WSG member. Overall, however, most workers felt that the organization of the WSG was very important in holding them together. Like Wiltz, both Esnard Gremillion and Ernie Rousselle stressed that if the WSG had not been organized, many wives, unaware of the issues in the dispute, would have put pressure on their husbands to return to work: "That's one of the best moves we made in the whole lockout," asserted Gremillion, "because looking at it in retrospect, if we didn't do that and they wasn't knowledgeable about what was going on, it would have made our job a hell of a lot harder too, trying to keep those guys from wanting to go back to work." As the company had locked out the entire bargaining unit, individual union members could not choose to return to work, as they could have in a strike. Workers could, however, have put pressure on their representatives to accept concessions. Rousselle credited the Monday night meetings and the organization of the WSG as the two most important factors in ensuring that he did not face pressure from local union members to accept the company's terms.[15]

Former WSG members also remembered how the lockout disrupted traditional gender relationships. Before the dispute, many wives had stayed at home while their husbands fulfilled the role of breadwinner. During the lockout, however, many men struggled to find well-paying jobs, and this pushed increasing numbers of women out to work, with some ending up earning more than their husbands. "A man, his identity is with being the breadwinner and there was a lot of women that were breadwinners during the lockout, women making more than their husbands," recalled Wiltz. This change put a strain on marriages, and WSG members blamed several divorces on the lockout. "A lot of people didn't handle it correctly," added Wiltz.[16]

The organization of the WSG did not help the small number of female BASF workers. Like the wives, these workers were reluctant to attend the male-dominated union meetings and felt rather isolated as a result. "I was one of the locked-out workers, but I still was not one of the guys," recalled Gladys Harvey, who had battled to secure a production position in the plant. One of only eight women workers, Harvey remembered that many locked-out union members met socially at each other's houses, but as a woman it was not easy for her to form the same kind of bond with individual male workers. At the same time, Harvey was also on "the other side of the table" from the WSG, which was a wives' organization that often referred to the locked-out workers as "our men." As the lockout wore on, however, she remembered that men were more accepting of her, and she consequently became a more active union member.[17]

Black workers made up around 15 percent of Local 4-620's membership, and they gave solid support to the union. Frank Smith, a senior black worker, emphasized that many of Local 4-620's members had been through the 1970 strike together, and they knew the importance of working as a group. "A lot of the guys that went through the lockout was some of the same guys that went through the strike, and we had experienced that before, and we knew, just like the strike came to an end, that this lockout would one day do that," he recalled. Other workers emphasized that the lockout cemented these bonds between union members. "It pulled everybody closer together," noted Putsy Braud.[18]

In the early 1980s, many African-Americans felt that BASF had not done enough to hire or promote blacks, and this helped to focus their anger against the company during the lockout. Just prior to the dispute, for example, several African-American workers filed complaints with the Equal Employment Opportunity Commission, asserting that a range of discriminatory practices still existed in the plant. "From my employment in 1976 there was an obvious attempt on the part of management to discredit me as an operator," noted one black worker. Several other workers, meanwhile, asserted that the plant had "a law for blacks and a separate law for whites."[19]

African-American workers also tended to live in Ascension Parish more than their white counterparts, and they repeatedly complained that the company should hire more local residents. Frank Smith, a parish resident, expressed how many felt: "We talked about local folk getting jobs before, and at one time they said that we didn't have enough qualified people in the area. . . . There are a lot of people here that tried to get jobs that did not get jobs. I can't say that it's because they're not a citizen here in Ascension Parish, but it is terrible that a man's got to drive sixty miles to come here and tear our highways up, the ones that we're paying taxes for, and take the money somewhere else and spend it, and here we've got people here working at Pay-Less for four ninety-five an hour. Then they say, 'Well, I don't have a job for them,' but you've got security, you've got grass-cutters, you've got all kinds of stuff. How much skill does it take for that?"

Like Smith, most black workers directed their frustrations at BASF and not the union. In its early days, Local 4-620 had clearly acquiesced in the maintenance of discriminatory job assignments, but by the time of the lockout black and white workers worked together well in the union, and several African-Americans held leadership positions. The most prominent was Leslie Vann, who served on the negotiating committee and held a variety of elected posts. Several other African-American workers, including maintenance mechanic Robert Washington, had also served on the negotiating committee in the decade before the lockout.[20]

The locked-out workers were not the only BASF employees who were put under pressure by the dispute. For the first three months of the lockout, salaried workers were required to run many hourly jobs. This put considerable strain on these workers, who had to work long hours in order to keep the plant running. In July 1984, Les Story, together with his plant managers, put together a detailed document that highlighted how the transfer of salaried personnel into production jobs was preventing them from carrying out many important tasks. Entitled "Assessment of What 'Lockout' Was Costing Plant," the document highlighted the pressure that the dispute placed on BASF's salaried workers before large numbers of new workers were hired. A detailed list of "work not being done" highlighted that "plant safety audits" and "planning" had both been sacrificed because of the dispute. In the area of safety, the document noted that there had been an "initial relaxation of programs." In the stores department, the manager reported that he was under severe pressure, with "problem periods" and "no backup for vacations, sickness, holidays." Overall, Story noted that the main emphasis was on "just keeping the systems running."[21]

Many of the salaried workers had relations who were locked out, and this threatened to cause some division within families. In most cases, however, union members understood the salaried workers' position and maintained personal friendships with them. "Basically the union knew that the salaried people were doing what they had to do," reflected salaried worker David Gilin.[22] The lockout did not witness the level of bitterness generated by many strikes in the 1980s, and there was almost no violence recorded during the dispute, mainly because union activists realized that it would only be used to discredit them.[23] Some salaried workers themselves sympathized with the union. "The Geismar management," recalled Richard Donaldson, "has always been divided about their personal feeling regarding the union and this situation was no different. We had managers who thought that the company had mistreated the union by imposing the lockout and we had managers who thought that the union got what it deserved by not coming to an agreement sooner."[24]

BASF executives, aware of the strains being exerted on their salaried workers, set about employing a new workforce. Shortly after the start of the lockout, the company brought in contractors to perform the 110 maintenance jobs, and in August they hired several contract labor companies to take over the operating work. The largest of these contractors was UMC, a local nonunion firm. BASF also, however, used union contractors during the lockout. Some members of craft unions, for example, had worked in the plant before the lockout, and they continued to do so during the dispute, reasoning that they were an independent union from the OCAW.

Dexter Guidry belonged to the pipefitters' union while he was working in the plant during the lockout. He emphasized that he followed the directions of his union's business managers, who told him to continue working. "Our business managers never told us not to," he recalled. "Maybe we should have, maybe they should have shown the solidarity with the OCAW, but they just never did. . . . We were doing this work before the lockout and I guess they felt we didn't actually cause the lockout." Relations between the union and nonunion contractors were frequently tense, with union contractors expressing more sympathy for the locked-out workers. Nevertheless, the failure of the craft unions to support the OCAW highlighted, not for the first time in the dispute, the lack of union solidarity that contributed to labor's problems in the 1980s. It also gave a powerful weapon to BASF, bolstering their claims that they were willing to employ union workers.[25]

Because of the economic recession affecting the chemical industry at the time, contract labor companies had few problems in finding workers willing to work at BASF. Louisiana's high unemployment rate was partly a reflection of the nationwide recession; by the end of 1982, national unemployment reached 10 percent, the highest rate since 1940, and it held at this level throughout 1983.[26] Although most of the products made at BASF were specialty chemicals that were not energy-intensive, many other chemical plants in Louisiana had traditionally produced energy-intensive, natural-gas-based products such as ammonia, urea, and methanol, and production of these chemicals was now hit by rising energy prices. By October 1983, employment in the state's chemical and allied products industry was down by 10 percent from the 1982 average of 33,800 jobs. Many of those laid off because of this downturn had registered with contractors in the hope of finding work.[27]

Many of the workers who took jobs at BASF had been laid off from plants within commuting distance of Geismar, with large contingents coming from the Kaiser and Ethyl plants in Baton Rouge. Others had previously worked at local Exxon, Allied Chemical, Uniroyal, Dow, and Stauffer plants. The company asserted that they were able to hire contract workers that had as much chemical plant experience as the locked-out workers. In June 1986, for example, BASF claimed that the new operators had an average of 10.5 years of work experience in the chemical industry, compared to 10.4 years among the OCAW workers they were replacing. Desperate for work, few of the workers hired were put off by the fact that the jobs on offer were only temporary.[28]

Among those hired by BASF was Myron Dedeaux, who had been laid off by the Ethyl Corporation in Baton Rouge in the fall of 1983. A Baton Rouge native, Dedeaux had worked at Ethyl for over ten years before his layoff, and he struggled to make ends meet after losing his job. When he was asked to work at BASF, Dedeaux admitted that he was tempted, but he also had doubts. At Ethyl, Dedeaux had been a member of the United Steelworkers' Union, and he hesitated to take a job at Geismar. "I didn't want to go across any union lines," he recalled. He eventually decided to enter the plant because the union was locked out and not on strike, ensuring that the new workers did not have to cross a picket line. "I didn't feel like I was a scab because I didn't cross any picket lines," he recalled. Even after sixteen years, however, he was still not completely at ease with his decision: "I always felt kind of guilty about it, and still to this day I do." Dedeaux, who was one of the first contract operators hired, remembered that many of his coworkers did have experience in the petrochemical industry, with a large contingent coming from Ethyl. Like Dedeaux, many of these workers had worked in unionized plants, although large numbers were also hired from nonunion operations.[29]

Three months into the lockout, the whole tenor of the dispute was changed when BASF announced to the union that it wanted to permanently subcontract the 110 maintenance jobs. Since June, the maintenance jobs had been performed by nonunion contractors, and the company claimed that this experience had convinced them that they could save money by permanently subcontracting the maintenance work. In August 1984, Les Story reported to Parsippany that the economic advantages of contract maintenance were "*compelling*," noting that open shop contractors were paid $11.53 an hour, compared to $14.72 for the union workforce. He estimated that the company could save $1.3 million every year by switching to an all-contractor maintenance workforce. Local managers had always known about these economic advantages, but they had stuck with the union's maintenance workers because of lingering doubts about the quality of a contract maintenance workforce. The experience of the lockout, however, convinced executives to make the switch. "Our concern for the effectiveness of a 100% contract maintenance force has been virtually eliminated," noted Story. "The two month lockout has proved to us that we can, and do, handle tough repair jobs with the same quality using contract mechanics." The company, Story asserted, was now using 75 contractors to carry out the work that 110 union maintenance workers had previously performed.[30]

Story also argued that contract workers offered other advantages. They were seen as more flexible, more willing to work longer hours and to take on a wide variety of jobs. For BASF, these advantages outweighed the drawbacks of using contractors. Managers noted that the contractors were "not as polished" as the union mechanics, and they acknowledged that contractor turnover would "always be higher" than among the union workforce. The company also admitted that it had experienced other problems with the contractors: "In the short period of time we have been operating with all contractor maintenance, we have experienced an unusual variety of maintenance problems." They asserted, however, that these problems were "compensated" by the greater enthusiasm of the contract maintenance workers and their "more efficient work habits." Story especially disliked the control that union maintenance workers maintained over their job assignments, complaining that they often refused overtime and had a narrow definition of the work they should perform. "It's just that your hands are kind of tied about what you can do," he recalled. "If a fella says, 'I don't want to do this work,' there's not much you can do to force him to do it, which was disconcerting to us." Acutely aware of the nonunion plants just next to BASF on the river, Story was keen to employ what he felt was a more flexible and competitive maintenance workforce.[31]

The union viewed BASF's decision to subcontract maintenance jobs as further evidence of the company's desire to destroy them. Most of the union's leaders, including President John Daigle and Chairperson Esnard Gremillion, were maintenance workers. The union therefore asserted that subcontracting the maintenance jobs would "wipe out" its most able activists. During the 1984 negotiations, 70 percent of the union's negotiating committee had indeed been drawn from maintenance workers. BASF, however, always insisted that its decision to subcontract the maintenance work was based purely on economic grounds, and removing union leadership was certainly never mentioned as a motive in company documents.[32]

In September 1984, BASF officials informed the union in negotiations of their intention to subcontract the maintenance work. The union, in response, insisted that the plant was only running with less personnel because contract workers were carrying out a great deal of overtime. OCAW negotiators asserted that the company would not save any money once it had paid for this overtime.[33] The company, however, argued that overtime was not "greater than normal" and reiterated that it was "more economical to run the operation with contractors." Unable to resolve these fundamental disagreements about the cost of contracting out the maintenance work, BASF implemented its decision to subcontract the jobs.[34]

In addition to this impasse over subcontracting, the two sides clashed over other issues. The union continued to refuse to agree to a company proposal that workers pay for a portion of health care costs, disputing BASF's arguments that these costs were rising. Although the two sides continued to discuss the proposed changes in seniority, they failed to make any progress. Management and labor also argued about wages, with BASF insisting that a small increase in wages would make the plant uncompetitive. Although the two sides met regularly under the auspices of a federal mediator, it was clear that a settlement was unlikely. In October 1984, for example, the company noted that there was a "vast canyon between the parties, which can be defined as a difference in philosophy and concepts, and until those differences are addressed chances for settlement are slim."[35]

During the first year of the lockout, Local 4-620's members engaged in a number of homegrown initiatives to try and further their cause. The local union, for example, initiated a letter-writing campaign in an effort to build support. Many wrote to state politicians, including Governor Edwin Edwards, to try and secure their support. In reply, Edwards repeatedly insisted that he had asked BASF to lift the lockout but his appeal had "fallen on deaf ears." He asserted that he did not have "the power" to settle the dispute on his own. Another leading state politician, Senator J. Bennett Johnston, meanwhile, responded to letters from union members and their wives by expressing his concern that the lockout had not been settled. "I will continue to monitor developments closely, and hope that this unfortunate situation can be resolved in the very near future," he noted. Beyond this, however, Johnston refused to intervene.[36]

Some locked-out workers and WSG members tried to secure the intervention of national political leaders. Patsy Cudd, the wife of a locked-out worker, pleaded with President Ronald Reagan to persuade the company to lift the lockout. Cudd had voted for Reagan in 1980 and 1984, the second time against her husband's wishes, and she hoped that the president would help the Geismar workers. "My husband says you are against organized labor and that nothing will come of my writing you," she noted. "Well my mother always taught me to believe the best of someone, so I'm thinking positive you will help us." Despite these pleas, Cudd's letter did not secure Reagan's intervention. Locked-out worker Kernest Lanoux questioned the company's tactics in a letter to Edwin Stenzel, the head of BASF in North America: "Is it really necessary to strip away the vestige of job security, the right to exercise seniority, the pride, the spirit, and the dignity from these experienced and qualified employees?" Stenzel apparently did not reply to Lanoux, although on other occasions he answered letters from the union by repeating BASF's arguments that it was not antiunion but was simply trying to improve the competitive position of the Geismar site.[37]

The fate of the letter-writing campaign illustrated the way that workers struggled to exert leverage against BASF during the first year of the lockout. Esnard Gremillion indeed later admitted that the union lacked a "formal plan of attack" during this first year and was unable to secure decisive help from political leaders.[38] After a year of the dispute, he admitted that union members were "worse off today than we were on June 15, 1984."[39]

The efforts of the international union to help the locked out workers in Geismar also went awry. A central part of the OCAW's strategy for the first year of the lockout was to work with trade unionists in Germany. BASF's German plants were represented by West Germany's chemical workers' trade union, Industrie Gewerkschaft Chemie-Papier-Keramik (I.G. Chemie), and OCAW leaders believed that I.G. Chemie leaders could influence the company to end the lockout, especially as the union had some apparent control over management decision-making. In 1951, the West German parliament passed a codetermination law for heavy industry, giving unions representation on companies' supervisory boards on an equal basis with shareholders. These supervisory boards had the power to appoint or dismiss management figures, and they could demand information on all company matters. The law also established that each board of directors must have at least one labor director, who was mainly concerned with working conditions. In 1976, codetermination was extended to other large-scale companies. This model of cooperation reflected the way that both management and labor felt a strong desire for peace and cooperation after the upheavals of the Nazi regime and World War II.[40]

These laws have helped to provide German unions with more stable membership levels than their U.S. counterparts. Between 1970 and 1984, a time when the membership levels of most American unions declined, those of German unions increased steadily.[41] Most unions belonged to the Deutscher Gewerkschaftsbund (German Trade Union Federation, or DGB), which was set up in 1949. In the mid-1980s, around 40 percent of workers were unionized. This percentage, while not high by European standards, was more than double the level of union membership in the United States. German unions also had much greater political influence than in the United States, with strong links to the Social Democratic Party (SPD), in particular.[42]

In the postwar era, organized labor made steady gains at the bargaining table and helped to win what one German historian has termed "the best working conditions for their members of any industrial society."[43] Led by Herman Rappe, himself an SPD member of the federal parliament, I.G. Chemie was a typically pragmatic union that had been successful at winning economic gains at the bargaining table without resorting to striking. By 1985, the union had 640,000 members and was the third largest in West Germany.[44] Although employers were reluctant to admit it, the co-determination laws helped to maintain industrial peace in West Germany and kept strike rates low. The low incidence of strikes was indeed one of the most eye-catching features of the West German industrial relations system. Some critics, however, charged that the laws robbed unions of their militancy, and OCAW leaders would themselves come to feel that I.G. Chemie lacked any real understanding of the kind of conflict that they were going through in Geismar.[45]

OCAW leaders were initially optimistic that I.G. Chemie could intervene and persuade BASF's German executives to put pressure on their American counterparts, with some claiming that they had received assurances that the dispute would be "straightened out."[46] The relationship between OCAW and I.G. Chemie deteriorated rapidly, however, after locked-out workers and their wives posted leaflets around the plant featuring swastikas and calling BASF "Co-Conspirators of the Third Reich." The flyers, which appeared in December 1984, added: "The nice people who helped to bring you World War II now operating with impunity in the '*United States of America*.' 'Union Busters' oppressing the constitutional liberties and lawful rights and freedoms of '*American workers*' in '*Geismar, Louisiana*.'" Using their spare time to research about the company's history, several locked-out workers and WSG members had read Joseph Borkin's *The Crime and Punishment of I.G. Farben*, which detailed BASF's involvement, as part of the I.G. Farben cartel, in the Holocaust. Workers felt that they had found a way of hitting back at the company.

CO-CONSPIRATORS
of
THE THIRD REICH

THE NICE PEOPLE WHO HELPED
to
BRING YOU WORLD WAR II
now
OPERATING WITH IMPUNITY
in
"THE UNITED STATES of AMERICA"

- - - "UNION BUSTERS" - - -

OPPRESSING
the
CONSTITUTIONAL LIBERTIES
and
LAWFUL RIGHTS and FREEDOMS
of

"AMERICAN WORKERS" in "GEISMAR LOUISIANA"

6. Swastika flyer distributed by local union, December 1984. (Courtesy Les Story)

"Whenever you get down to where you're getting the shit beat out of you, you're going to use whatever you can to take care of yourself," recalled one. "We just dug up their history, and the history of this company goes straight back to the death camps in World War II."[47]

At a subsequent demonstration outside the plant led by the WSG, at least one protester publicly carried a sign displaying swastika emblems with lines through them next to the slogan "Foreign Oppression of American Workers." A photograph taken by Les Story shows OCAW's Ernie Rousselle (with his back to the camera) talking to the protester. In a retrospective interview, Rousselle acknowledged that he gave his approval for the sign to be used. "I was aware of it before it went on and I allowed it to go on," he admitted. While accepting that his decision may have been a "mistake," Rousselle felt that BASF's involvement with the Third Reich was undisputed history and was a legitimate weapon for the union to use. Many of the locked-out workers shared his opinion, remaining keen to link the company's lockout in Geismar to its war record. In a letter written to the local press, for example, one worker called the company's actions "pure fascist." One salaried worker also alleged that he received a telephone call where an unidentified caller had stated that he was "a Nazi"

7. Ernie Rousselle talking to a female protester with a swastika sign. (Courtesy Les Story)

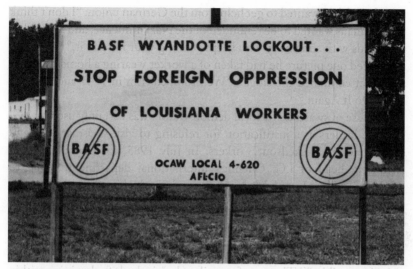

8. "Stop Foreign Oppression" billboard, circa December 1984. (Courtesy PACE International Union)

who was "betraying" his country. Other contract workers reported that the Nazi leaflets were left on their cars.[48]

Local union president John Daigle also supported the swastika leaflets. They had, he asserted, raised the morale of the locked-out workers and had brought up the legitimate issue of "a foreign country taking advantage of local people." The company claimed that two thousand swastika flyers were ultimately distributed and that some workers displayed them in their cars. The local union also placed a large billboard overlooking Interstate 10, which ran from Baton Rouge to New Orleans, that read "Stop Foreign Oppression of Louisiana Workers." The billboard was placed on land owned by a locked-out worker, a site that would later be used to publicize several other union allegations. The locked-out workers also erected several other "Stop Foreign Oppression" signs in the area.[49]

There is no doubt, however, that the use of swastikas helped to destroy the union's strategy of working closely with its German counterparts. BASF officials themselves helped to widen the split between the two unions, with Les Story admitting that he played an active role in ensuring that the company's German headquarters knew about OCAW's use of swastikas. "I sent them copies of the leaflets that they put on the telephone polls," he later acknowledged. Story was so keen to publicize these leaflets because he sensed that the union's use of swastikas was a "huge tactical mistake" if they wanted to get help from the German union: "I don't think the Germans wanted to be reminded of the Nazi influence, and it's hard to be friends with somebody who keeps rubbing your nose in it." Story also publicized one picture he had taken of a worker wearing a home-made T-shirt that contained the slogan "We Stomped 'Em in 1945 and By God We'll Do It Again."[50]

The use of swastikas by local demonstrators was certainly used by I.G. Chemie leaders as a justification for refusing to visit Geismar and lend support to the locked-out workers. In July 1985, for example, Bernd Leibfried, head of I.G. Chemie's international department, wrote to OCAW president Joseph Misbrener to explain why his union had recently canceled a trip by a delegation to Geismar. "This demonstration with swastikas was and is the *only* reason why we could not send an I.G. Chemie delegation to Geismar," he fumed. A telegram sent to OCAW headquarters from I.G. Chemie also noted that the banners with swastikas made a visit "impossible."[51] The use of swastikas decisively shifted opinion within I.G. Chemie against supporting the Geismar workers. Relatives of several I.G. Chemie leaders had been killed by the Nazis in concentration camps and they felt particularly strongly that the OCAW should not be supported. OCAW leaders, however, felt let down by the German union, feeling that they should have done more to support the workers in Geismar. OCAW vice-president Robert Wages, for example, wrote that "I.G. Chemie's cancellation makes a mockery of the notion of international trade union solidarity." Wages' letter, however, only made the Germans angrier.[52]

In correspondence with BASF executives, who were shocked and alienated by the union's tactics, OCAW leaders sought to distance themselves from the swastikas, insisting that they had not been sanctioned by the international union. This was not completely accurate, however, as Rousselle, an international representative, had consented to their use. While the international union was keen to prevent the further use of swastikas, it was nevertheless quite willing to use the company's war record against it, sensing that it could help build support for the union's case within the United States. Union leaders argued that this war record highlighted that BASF was a company with a long history of treating workers unfairly. In the early part of 1985, in fact, several union press releases attacked BASF's wartime record. The union claimed that the company's "ancestral history is one that cannot be viewed with patriotic pride" adding that "in many ways BASF's attack on the American worker appears to be an extension of the dreaded war between the two nations that ended four decades ago."[53] These allegations were repeated in subsequent union publications. On two occasions in 1985, for example, the *OCAW Reporter*, the union's official newspaper, detailed BASF's war record, with one article asserting that the company's conduct in Geismar reflected "a firmly established heritage . . . born and nurtured under Nazi fascism."[54]

Throughout the dispute, in fact, the union would repeat this line of argument, thus continuing to alienate the German union and the company, which both considered OCAW's approach to be a "slur." In 1988, a video produced by the union to publicize the dispute gave considerable attention to BASF's war record. Within the United States, these tactics did help to secure support for the union, especially from church leaders who came to view the lockout as part of a broader pattern of immoral behavior. The video was also distributed in Germany, however, where it alienated I.G. Chemie still further. OCAW leaders were unrepentant. "They were very offended by our using its history against them, but we did it anyway," recalled Robert Wages.[55]

Following the union's unsuccessful efforts to enlist the support of I.G. Chemie, the two sides returned to the bargaining table. BASF, sensing that it had the union on the ropes, pressed for a settlement, telling OCAW negotiators that their efforts to enlist the support of I.G. Chemie had "accomplished nothing" and that they therefore lacked leverage. OCAW representatives felt, however, that the company was openly trying to break them and that they had to make a stand against concessionary bargaining. Both sides refused to budge from their fixed positions, with the contracting out of maintenance and seniority continuing to be the main stumbling blocks.[56]

After a year of the dispute, local 4-620's members were as determined as ever, but they were also conscious that they were failing to exert any real leverage against BASF. In April 1985, for example, Esnard Gremillion told the *New Orleans Times-Picayune* that union members were struggling to survive. "They're controlling our whole life," he commented. "They've got us cornered and they're going to keep putting the pressure on until we break." Both OCAW leaders and local union members knew that if they were to have any chance of turning their fortunes around, they needed to try some new tactics.[57]

Bhopal on the Bayou?

For the first year of the BASF lockout, the international union had aimed to use its contacts with German trade unionists as a way of exerting pressure on the company. Following what OCAW leaders referred to as "the collapse of our relationship with I.G. Chemie," they decided to launch their own corporate campaign against BASF.[1] The top leadership of the OCAW took this decision because they felt that a great deal was at stake in their fight with the German chemical giant. A former oil refinery worker from California who was elected OCAW president shortly before the start of the lockout, Joseph Misbrener felt that his union had to make a principled stand against BASF's demands in order to retain its broader credibility as an organization. If they had simply given in, he wrote BASF chairman Hans Albers, it would have set a "dangerous precedent" that other companies could have followed.[2] The union was, however, clearly struggling to exert any pressure at all against BASF, and Richard Leonard, the staffer assigned to head the campaign, admitted that he was initially reluctant to take on such a powerful company. Nevertheless, OCAW leaders realized that they had to find new ways of fighting back, and they pushed him to look into the dispute. As he later recalled, the union's leaders told him, "Look, get your ass to Louisiana."[3]

The OCAW was not the first union to launch a corporate campaign. As early as the 1970s, some unions, feeling that strikes were becoming increasingly ineffective, looked for other ways to exert pressure against companies that vehemently opposed organized labor. One of the earliest and most visible corporate campaigns was launched by the Amalgamated Clothing and Textile Workers' Union (ACTWU) as part of their ongoing struggle to organize J. P. Stevens. Facing an employer who had successfully resisted traditional organizing techniques, ACTWU staffer Ray Rogers pioneered new tactics. Rogers highlighted the fact that several J. P. Stevens board members also sat on the board of other companies, successfully connecting executives such as the chairman of Avon cosmetics, David W. Mitchell, with the textile giant. Targeting women who used Avon products, Rogers publicized Stevens' opposition to the unionization of its women employees and helped pressure Mitchell into resigning from the Stevens board. ACTWU also urged the public to boycott J. P. Stevens products as part of its campaign. The campaign attracted a great deal of negative publicity for J. P. Stevens, influencing the firm to sign its first contract with a union in 1980.[4]

The victory at J. P. Stevens produced a great deal of optimism within the labor movement, with many union leaders feeling that they had discovered the new technique that they needed to reestablish union power. In the early 1980s, however, several corporate campaigns failed. Rogers himself found it impossible to repeat his success when he was hired as a union consultant by the United Food and Commercial Workers (UFCW) in 1984. The UFCW's high-profile strike against Hormel Meatpacking Company ended disastrously when the international union withdrew its support for the local union. Although Rogers continued the campaign for many months, he was unable to bring the company back to the bargaining table. The Hormel failure showed that the corporate campaign was not a quick fix for America's ailing union movement. Other disputes highlighted the fact that determined employers like Hormel could successfully resist corporate campaigns and showed that unions needed to use a wide variety of tactics in order to stand any chance of success.[5]

In the 1980s, the OCAW also turned to the corporate campaign as an alternative to striking, especially as the high proportion of salaried workers in the chemical industry made it possible for companies to run plants even when production workers did walk out. Shortly after buying a refinery in Memphis, MAPCO, an Oklahoma oil and gas company, demanded major concessions from the OCAW-represented workforce. With negotiations reaching an impasse in January 1984, the local union, with the help of the international, avoided striking and instead stayed in the plant. The OCAW launched its first corporate campaign, pioneering many of the tactics that would later be used against BASF. The local union linked up with community groups to bring pressure against the refinery for air quality violations, the illegal burial of lead sludge, and contamination of the groundwater. OCAW also voiced residents' concerns about the safety of the refinery, as well as charging that MAPCO had unauthorized use of city water in its cooling towers. As in the BASF campaign, union members, attempting to encourage shareholders to pressure MAPCO to settle the dispute, attended the company's annual meeting en masse. In May 1985, the MAPCO campaign did end successfully for the union, as the company agreed to a new contract that maintained most of the provisions of the earlier agreement, as well as granting workers a wage increase.[6]

The MAPCO campaign was the brainchild of Richard Leonard, a staffer at the OCAW's Denver headquarters. Born in 1945, Leonard had left his home in upstate New York to attend the University of Colorado in 1963. After obtaining a degree in economics, Leonard was attracted to working in the union movement and was hired by the OCAW, which was based in Denver. While working in research and education, Leonard earned a reputation for his dedication and ability to provide thorough, detailed research on companies that the union bargained with. In 1983, he began to work full-time on special projects for the union, and for the rest of the decade he was to run a series of corporate campaigns, with the majority of his time taken up by the BASF dispute. In his campaigns, Leonard repeatedly reached out to the communities around chemical plants, asserting that they shared a common interest with workers. "If the company is beating up its own employees, they're probably beating up everybody else too," he charged. "You make toxic chemicals, if they're going to take short-cuts, it's not only going to be with the workers, it's going to be with everybody else." Leonard realized that the community was potentially a powerful ally for unions: "If there's a fight over jobs and the environment, you really want the community as an ally and not an enemy, because I think ultimately the community will prevail in an issue like that. You want the community to see the workers as important. You want the community to visualize the union as a necessity to allow workers to be able to speak up concerning these environmental or health and safety questions and be the eyes and ears of the community, ideally."[7]

Following the MAPCO fight, Leonard applied his ideas in several other campaigns. Just prior to the BASF lockout, for example, he drafted a letter to the shareholders of the drug manufacturer R. P. Scherer in an effort to try and persuade the company to recognize an OCAW local union in North Carolina.[8] During the early stages of the BASF lockout, the union also launched a campaign against Minnesota Mining and Manufacturing (3M) after the company decided to close its audiotape- and videotape-making plant in Freehold, New Jersey. The OCAW, which represented the 360 workers at the plant, asked 3M to keep the facility open. When the company refused, the union enlisted the help of rock star Bruce Springsteen, who had grown up in Freehold, to launch a campaign against plant closures. The union was keen to work with Springsteen because he had sung about the devastation caused by the 1964 closure of a Freehold textile plant on his hit recording "My Hometown." After meeting with rank-and-file members, Springsteen placed his name on full-page newspaper advertisements that implored "3M: Don't Abandon Our Hometown!" Other advertisements pointed out that some of the biggest users of 3M products were the recording companies. The campaign was also supported by Willie Nelson and became the subject of a report shown on *20/20*, ABC's current-affairs show. The union's efforts did not stop the plant closing, but they did highlight to industry that the OCAW was willing to use imaginative tactics in order to make its point.[9]

When he was assigned to head the BASF campaign, therefore, Leonard already had a great deal of relevant experience. On an initial trip to Louisiana, one of his first tasks was to try and convince Ernie Rousselle not to sanction the further use of swastikas. "The use of swastikas, etc. could be very self-defeating at this time," Leonard warned. It would discredit the union, he argued, and mark it out as a "fringe" element in the public's mind. Leonard recognized that the symbols united the locked-out workers and expressed their frustration with the company, but he was aware that a successful strategy also required the mobilization of broader public opinion behind the union. He intended to do this by isolating BASF. "It is important in our opinion to establish the union as the responsible party and the company as the irresponsible party," he wrote. "I would like to try to establish the company as a 'Corporate Rogue Elephant,' loose in the community, polluting, cheating on taxes, trampling on workers and the public in general. I don't we can establish this view of the union if we're out there talking about swastikas, Nazis, etc. The use of these symbols would be fairly last ditch and I don't think we're this desperate." Leonard thus mapped out the direction that the campaign was to move in, and he succeeded in preventing further circulations of swastikas.[10]

After talking to a wide variety of workers, Leonard realized that they possessed a close knowledge of the company's environmental practices that could be used to embarrass BASF. "We became very enthusiastic about it in terms of all of these various degrees of company misbehavior in the Geismar area," he recalled. "And we particularly noted some problems in the environmental area."[11] Leonard obtained a large engineering map of the Geismar facility and supplied workers with large colored pens. At meetings, they then came up to the map and drew circles in areas where they knew environmental contamination had taken place. As well as providing a way of exerting leverage against BASF, an environmental campaign, Leonard realized, would provide a constructive outlet for the energies of the rank and file. When he returned to Denver to map out strategies for the corporate campaign, he identified "community and worker health and safety" as key issues.[12]

Leonard had realized that the environmental issue had a great deal of potential, and he was keen to hire a specialist who could explore it further. It was especially important to find an activist who could be based in Louisiana on a full-time basis, as Leonard worked out of the OCAW's headquarters in Denver and needed to continue directing other campaigns in addition to the BASF struggle. In the 1970s, Leonard had become close friends with OCAW international vice-president Tony Mazzochi, and he closely supported the veteran leader's unsuccessful efforts to win the OCAW presidency in 1979 and 1981. Around the time that Leonard was drawing up strategies for the campaign, he called his old friend, who was now working for the Labor Institute, a New York–based institution that taught labor relations and economics classes to rank-and-file union members. Leonard asked Mazzochi if he could recommend a strategist who could go to Louisiana and work on the campaign. Mazzochi immediately suggested Richard Miller, a young activist who had been working at the institute for the past two years and whose intelligence and energy had impressed him. For the next five years, Miller was to be the key figure in the union's environmental work.[13]

Born in Boston in 1954, Miller worked in a wide variety of jobs after graduating from high school, including a long spell as an auto mechanic. Never happy "punching a clock," he was drawn to the University of Massachusetts in Amherst, where he took a degree in labor studies. This course attracted him to the labor movement, and after graduating he went to work for the Labor Institute. In the fall of 1985, Miller agreed to travel to Louisiana to work on the BASF campaign, believing that it would only be a temporary assignment: "Mazzochi said, 'Look, get down to Louisiana. Dick Leonard's there in this campaign. It will take you six months. It will be a long struggle'—he meant six months—'you'll learn a lot. You'll bring it back to the Labor Institute.' I said, 'Okay,' and so five and a half years later, I was there, still."[14]

As a Jewish northerner who had never been to the South before, Miller recalled that his early days in Louisiana were a massive "culture shock." Holed up in a motel on Airline Highway in Gonzales with few contacts in the community, Miller found his first weeks in the Pelican State lonely. He was, however, an incredibly energetic activist who began to work long hours investigating BASF and ways that the union could exert leverage against it. Like Leonard, he realized that the union could use BASF's environmental record as an effective way of putting economic pressure on the company. It was, he later reflected, the company's "Achilles' heel."[15]

Those who worked with Miller remembered his energy and total commitment to the campaign. Marylee Orr, an environmental activist who was to become one of Miller's closest allies, described him as "an extraordinary talent," while state official Willie Fontenot viewed him as "one of the smartest guys I've ever worked with." Described by Fontenot as "one hundred and twenty-six pounds soaking wet," Miller was a thin, live-wire of a man whose dedication began to impress the locked-out workers. His work rate quickly became legendary, with many fondly recalling late-night strategy sessions in his motel room, which became the informal base for the emerging campaign. Within a short time, Local 4-620's members had grown accustomed to Miller's idiosyncratic work habits and fully trusted him. "Richard Miller, I just can't say enough about him," noted Bobby Schneider. "Workaholic, don't never sleep, no concept of time. He might call you at three o'clock in the morning because . . . . the idea hits him and you caught an idea. You got to the phone and there was Richard and he's wide open, and he's, 'We've got to do this and this is what you've got to do on it,' and there's no doubt Richard was the spark for getting the environmental issues [going]."[16]

As they investigated BASF together, Miller and Leonard realized that the Geismar plant manufactured chemicals "similar in many respects" to those made by Union Carbide in Bhopal, India.[17] On December 3, 1984, more than 3,500 people were killed, and thousands injured, by a leak of the toxic gas methyl isocyanate at Union Carbide's Bhopal plant. The Bhopal accident was history's worst industrial disaster, and it was reported extensively in the American press. The company claimed that the accident had been caused by a disgruntled worker who had sabotaged equipment, while the Indian government charged that the Bhopal plant was badly maintained and had design flaws. As the two sides pressed their conflicting claims, the Bhopal story continued to make the headlines long after the disaster. It took until February 1989 before the case was finally settled by the Indian Supreme Court, which ordered Union Carbide to pay $470 million in damages to the Indian government.[18]

The Bhopal disaster focused public attention on the issue of chemical plant safety and put the industry on the defensive. *Chemical Week* urged companies to support efforts to improve safety in the wake of Bhopal, noting that it was essential to try and win back the public confidence lost as a result of the accident.[19] As public pressure for tighter environmental laws mounted, the industry journal worried that corporations would "soon feel a special bite from Congress because of the accident in India."[20] In the aftermath of the tragedy, several members of Congress pushed for tougher environmental laws, including Michigan Democrat John Conyers, Jr., chairman of the House Judiciary Committee's Criminal Justice Subcommittee. Aided by California Democrat George Miller, Conyers proposed a corporate responsibility law that made it a crime for companies to knowingly conceal dangerous products or working conditions that could result in an accident.[21]

In August 1985, moreover, a serious leak from a Union Carbide plant in Institute, West Virginia, sent 135 people to the hospital. The accident was not, as the *New York Times* pointed out, the most serious chemical accident in American history, but with Bhopal a recent memory, it produced a great deal of public concern. The *Times* noted that the accident had dealt a serious blow to the credibility of the chemical industry, especially as it occurred just after executives had been stressing the measures that they had taken to ensure that their plants were safe.[22]

While the issue of chemical plant safety was being debated in the press, both Leonard and Miller heard allegations that the safety of the Geismar plant had suffered since the company had locked out the OCAW. As early as January 1985, for example, Ernie Rousselle had claimed that BASF was using contract employees with "no experience whatsoever" to run the plant. That month, a serious accident had occurred after a worker loaded chemical residue into a dirty truck tank, leading to a citation from the state government.[23] The union also obtained other documents which indicated that the accident rate had gone up significantly since the lockout had begun. Many of these documents were passed to Miller by sympathetic salaried employees, some of whom were related to locked-out workers, while others were given to the union by unionized contractors who were working in the plant. "People had family members working in there, still," recalled Miller. "Gremillion's son was working in there while he was locked out. There was a lot of people who were in there, and so paper was still flowing out of the plant, and so we had all these documents."[24]

After collecting a great deal of data, Leonard and Miller decided to publicly question the safety of the Geismar plant at simultaneous press conferences in Bonn, West Germany, and Washington, D.C. The two activists also linked the BASF plant to the Indian tragedy with the phrase that they had used to publicize the conferences. Pointing out that the Geismar plant manufactured many of the same chemicals as Bhopal, they questioned whether there could be a "Bhopal on the Bayou." The phrase would subsequently be used by the union throughout the dispute. The AFL-CIO's Industrial Union Department, set up in 1955 to assist individual unions in running campaigns against companies, helped to publicize the conferences, which were planned to coincide with the first anniversary of the Bhopal accident.[25]

While in Germany seeking the support of I.G. Chemie, Leonard had made contact with some leaders of the German Green Party. Emerging out of the protest movements of the 1960s, the Green Party had grown steadily in the 1970s and early 1980s, reflecting the upsurge in environmental consciousness taking place across Europe. After seeing some of the documents that the union had obtained, the German environmentalists became concerned about the safety of the plant and agreed to lend their support to Leonard's campaign. To coincide with the union's press conferences, the German Green Party issued a statement condemning the lockout as "an outright scandal." "In light of the dreadful catastrophe in Bhopal, India, one year ago," they added, "this neglect of security standards and of the safety of the workers represents a highly irresponsible business policy."[26]

In flyers distributed at the press conferences, the union claimed that the Geismar plant was operated by poorly trained temporary workers who were "exposed to releases of extremely toxic chemicals on a daily basis." To support its case, the union used company sources to cite two accidents that had occurred in the plant during the lockout, including a major phosgene leak that required seven workers to be hospitalized. Phosgene, which had been employed as a chemical weapon during World War I, was the most dangerous chemical used in Geismar; if breathed in, it reacts with the moisture in the human body and forms hydrochloric acid, filling up the lungs with water. Prolonged exposure could be fatal. Other workers had been exposed to toluene diisocyanate (TDI), which was used to blow urethane foam. Not as dangerous as phosgene (which was itself used to make TDI), TDI was still a mutagen and embryotoxin. Asserting that the plant was "out of control," the union also stressed the high turnover rate among the temporary workers, which, according to company documents, was in excess of 50 percent throughout 1985. The union's concerns were shared by Dr. Karim Ahmed, a senior staff scientist at the Natural Resources Defense Council, and by Dr. James Melius, a director of the National Institute of Occupational Safety and Health. Ahmed, in particular, emphasized the similarities between the Bhopal and Geismar plants, including "woefully inadequate training, high worker turnover rates and maintenance cutbacks."[27]

By linking the dispute to the Bhopal accident, the union sought to capitalize on public and media concern about chemical plant safety. The tactic worked. The New York Times, which had failed to mention the BASF lockout until this point, ran a detailed story about the union's allegations, adopting the headline "A Bhopal on the Bayou?" The article outlined the dispute and presented the union's case that the plant was unsafe. The Wall Street Journal detailed the union's allegations in a front-page article headlined "Is It Safe?" Although the company denied the allegations, the bulk of the coverage in both papers concentrated on the union's case, much to the annoyance of BASF. By raising the issue of plant safety, the union had secured national media attention, and the focus of coverage had shifted away from traditional labor-management issues and toward environmental issues.[28]

Within Louisiana, the union used its billboard situated next to Interstate 10 in Baton Rouge to publicize its allegations. The new "Bhopal" billboard replaced the "Stop Foreign Oppression of Louisiana Workers" sign that local union members had previously erected, indicating the shift in focus that Leonard and Miller had brought to the union's campaign. Every day, thousands of motorists drove past the billboard, and over a decade later many local people still remembered it. It was, noted one environmental activist, "a tremendous statement." The sign helped to increase the profile of the dispute within the state, and press coverage began to be much more extensive than before. The Baton Rouge Morning Advocate, which tended to side with BASF during the dispute, even argued that the sign should be taken down because it could hinder the state's efforts to attract both industry and tourists.[29]

The Bhopal billboard undoubtedly worried BASF management. "It concerned them locally, from a community relations standpoint," admitted Bill Jenkins. Jenkins also felt that the sign "got the attention of the parent company," claiming that "it was a hell of a billboard." The union's

9. The "Bhopal" billboard. (Courtesy PACE International Union)

earlier "Foreign Oppression" sign made no mention of environmental issues and was of less concern to the company than the Bhopal billboard. Correspondence between BASF executives and their public relations consultants indicate that the company wanted a "repositioning of the Geismar situation as a labor/management struggle rather than a health and safety issue." As both Leonard and Miller realized, health and safety issues made the company uncomfortable and increased the union's ability to generate public support.[30]

Local residents, fearing that they could be killed or injured by a chemical release, were especially concerned about the union's allegations. St. Gabriel resident Daisy Goodlow, for example, wrote Les Story looking for reassurance that the plant was safe. She warned BASF that they were "gambling with human lives" and asked the company to address the union's claims that a Bhopal-type accident could occur in Geismar. "I, with others, are looking forward to an honest answer," she concluded. The Ascension Parish Police Jury (a police jury being a Louisiana parish's governing body) also wrote Story to express its members' concerns about the safety of the plant. "There are reports of breaches of fundamental safety rules by contract employees which have led to injury and exposure to toxic fumes," the police jury's letter noted. "Since the BASF plant uses and produces the same range of toxic and highly reactive chemicals produced at the ill-fated Bhopal, India plant, the employment of temporary workers raises concerns for community welfare." In the wake of the Bhopal accident, the police jury urged BASF to keep local people better informed about accident risks. Story responded to these letters by reassuring residents that the plant was safe, although he refused to divulge detailed information about the company's operations.[31]

Although both Leonard and Miller were pleased with the headlines they had secured, they were keen to exert sustained pressure on BASF. The union therefore submitted its allegations to Congressman Conyers, who wanted a $1 million fine for corporations if they misrepresented or failed to disclose health and safety information affecting workers and surrounding communities. The BASF plant was one of twenty plants that were being investigated by Conyers' subcommittee, but was the only one in Louisiana. Conyers was not ultimately successful in turning his proposals into legislation, but the efforts of his subcommittee produced a great deal of concern among industry leaders.[32]

Miller and Leonard worked tirelessly to produce two detailed reports that they submitted to Conyers' subcommittee. As well as asserting that the plant was dangerous, the union charged that BASF was engaged in "herculean efforts" to conceal and misrepresent the degree of danger to the public.[33] The reports cited many company documents to support their claims, reproducing many of them in appendices to validate their authenticity. One memo written by a safety manager for BASF's largest contractor workforce discussed the need "to monitor for air contaminants in the work areas." It added: "The personal protective equipment used by operators at BASF must be improved to prevent exposure to TDI and phosgene. The large number of chemical exposures suggests the need to carefully evaluate the situation and develop control methods. . . . These problems point to the lack of personal protective equipment, as well as inadequate training and engineering controls." Other documents highlighted the failure to provide proper respiratory protection to workers, alarm failures, and the accidental release of chemicals, some of which were shown to have migrated off-site. Other evidence suggested that BASF had not used preemployment physical examinations, including respirator fitness examinations, for contract employees, even though such examinations had always been given to OCAW workers before the lockout.[34]

Company data also showed that the number of worker injuries requiring medical attention had increased by 150 percent during the lockout. The union linked the high accident rate at the plant to high turnover, citing company data that showed that 60 percent of all injuries occurred to workers during their first sixty days of work. The union also drew out the comparisons with the Bhopal plant, claiming that management at the Indian plant had also made extensive use of "improperly trained temporary contract workers" and had sought to downplay the dangers that the plant had represented to the public. "Bluntly put," the report concluded, "BASF Geismar is a Bhopal in the making."[35]

The union's allegations clearly concerned the subcommittee. Worried by the comparisons with Bhopal, Conyers' aide Julian Epstein claimed that the Geismar plant was "being operated in a highly dangerous and even reckless manner. We are very concerned that BASF represents a ticking toxic timebomb that is holding both the community and the workers hostage."[36]

The threat of a congressional investigation into the safety of the plant clearly worried BASF's management. Les Story recalled that the reports that the union submitted to the House Judiciary Committee's criminal justice subcommittee made the company more "nervous" than any other aspect of the corporate campaign: "They went to the U.S. Congress to complain about our safety record, and the last thing in the world you want to have is a congressional investigation of anything. It's a widely publicized thing, it can easily get distorted, people have all sorts of political agendas, and so we spent quite a bit of effort and quite a bit of money to show the U.S. Congress that there was no basis for this allegation." Story himself visited Washington on several occasions to meet with legislators and convince them that the plant was safe, while the company's public relations consultants closely monitored reactions to Conyers' proposals on Capitol Hill.[37]

Other business leaders also worried about the investigation. Lee Griffin, chairman of the Baton Rouge Chamber of Commerce, wrote Louisiana Senator J. Bennett Johnston to express his opposition to Conyers' plan. Griffin worried that the investigation would drive away business and cost the state jobs. Louisiana's "perceived labor climate" would be damaged, he claimed. The union's accusations also concerned state political leaders. In February 1986, Louisiana senator Russell Long, for example, wrote of his concern about the growing "tension" between the union and BASF. Calling the company "an asset to the economy in and around Ascension Parish," Long hoped that the union's allegations would not deter BASF from investing in the state. Afraid that the lengthy dispute was damaging the state's image and its efforts to attract industry, Long joined Governor Edwards in supporting efforts to mediate the dispute.[38]

BASF also produced a detailed and glossy document that it submitted to the House Subcommittee on Criminal Justice in order to rebut the need for a congressional investigation. The company introduced its ninety-six-page document by asserting that the union's safety allegations were "a conglomeration of distortions, exaggerations, statistical misrepresentations and factual errors."[39] The company argued that the Geismar plant was a "modern, state-of-the-art chemical plant" that was "one of the safest" in the nation. The plant in Bhopal, BASF pointed out, did not possess such "state-of-the-art" equipment. BASF claimed that even if a toxic leak occurred, it had a sophisticated community emergency alert system. The company insisted that its high safety standards had not been jeopardized by the hiring of temporary contract workers, asserting that the recession in Louisiana had allowed the company to hire "exceedingly well-qualified" workers who had been laid off from other chemical plants.[40]

While it certainly mounted a spirited defense of its safety record, the company was not able to completely dismiss the union's case. BASF did admit that contract workers had higher accident and turnover rates than the OCAW members they had replaced, although they disputed the union figures. The company sought to publicly downplay the significance of rising turnover, asserting that it primarily occurred in the least-skilled jobs and had "little operational significance." The company also accepted that a degree of danger was inevitable when chemicals such as phosgene were used, but it argued that this danger was minimized because few people lived near the plant. This was a contrast to Bhopal, where "many of the victims were literally at the gates of the factory." In describing the larger community of Gonzales as the nearest "town," however, BASF officials conveniently overlooked the residents of Geismar, some of whom did indeed live very close to the plant.[41]

Looking back on the dispute, Story stressed that the union's allegations did have some validity, especially in the early stages of the dispute. "The union made what I would call legitimate attacks on the safety record of the company," he acknowledged. ". . . Someone who was coming in who had never worked in the plant before wouldn't be as safe as the person who had worked there for years before. In the early days, that was a fact. . . . Admittedly, there was probably no one in the area that had a plant as safe as ours when the OCAW people were there." He stressed, however, that the union's allegations became less and less accurate over time: "The safety record continued to improve, so I think in the beginning it was legitimate, but as time went along, it had no meaning at all."[42]

Although BASF publicly denied that there had been a decline in safety standards, Story's own records confirm that there were real problems, particularly in the early stages of the lockout. Data from the first eighteen months of the lockout highlighted that reported safety-related incidents had increased steadily over the period. "For what it's worth," one manager related to Story, "this shows a steady increase in reported incidents." In 1985, for example, there were 338 reported incidents compared to only 223 in 1983 and 170 in 1982.[43] Files show records of several workers' having been hospitalized in the first year of the lockout after exposure to chemicals such as chlorine and phosgene.[44] Although the company publicly sought to downplay the higher accident rate among contractors, in March 1985 Story did write to the managers of the contract companies warning them that he wanted "a substantial improvement in safe performance from our contractors in 1985."[45]

Workers who were employed in the plant during the lockout also felt that there were safety problems. Dexter Guidry, a contractor working for National Maintenance, remembered that he was concerned about the plant's safety record during the dispute. "BASF was not real conscientious as far as trying to protect the community," he asserted. "I mean I can remember working here for National. They're called effluent ditches, which is a ditch that goes out towards the river. There were nights when it was black. Whatever they were putting into it, and we'd work in it with some pumps and try to pump it back, but that stuff was going to the river. So I know they were dumping some bad stuff."[46]

Although the union's tactics caused concern, they also led to some support for BASF. While Guidry was a union contractor, nonunion contractors and salaried employees often took the OCAW's allegations personally and hit back at the union. One president of a contractor company, for example, claimed that the BASF plant was a "well-maintained and safely operated facility."[47] A UMC operator, meanwhile, objected to the union's characterization of contract workers as a "safety hazard" and claimed that they were as "proud" of their work as union members were.[48] John A. "Johnny" Berthelot, the mayor of Gonzales, was a consistent supporter of the German chemical-maker, calling the company "a good neighbor" and noting that the chemical plants were "the lifeblood of our area."[49] Throughout the dispute, Berthelot worried that the union's campaign would cost his town investment and jobs. Support was also expressed by the Greater Baton Rouge Chamber of Commerce, which was anxious to assure the company's German management "that Louisiana and the Baton Rouge area offer continuing attractive capital investment opportunities for BASF Corporation."[50] The company took heart in this support against the union's "smear campaign," and the support influenced the company to stand firm.[51]

Leonard had always realized that the campaign would need to exert pressure against BASF in a wide variety of forums in order to be successful. As part of his effort to create international pressure against the company, he urged the Organization for Economic Cooperation and Development (OECD) to investigate allegations that BASF had violated several of the OECD's guidelines for multinational enterprises. Under the guidelines, the parent company was required to treat workers in foreign subsidiaries in a similar fashion to workers in its home state. As offensive lockouts were illegal in West Germany, the union claimed that BASF was treating its workers in Louisiana too harshly.[52]

The union began by seeking the support of the U.S. Department of State in filing charges against the company with the OECD. When that effort failed, with the State Department refusing to intervene in an industrial relations dispute, the union acted on its own and contacted the OECD's Trade Union Advisory Committee (TUAC) in Europe. The TUAC reported the complaint, requiring company executives to defend their case. The OCAW did not, however, achieve its goal that either TUAC or the OECD as a whole condemn BASF. Despite this failure, Leonard asserted that the union's efforts were still "driving the company nuts." U.S. managers admitted, indeed, that the union's approach to the OECD worried executives in Germany. "The headquarters, the parent company, was concerned that we appeared before the OECD," acknowledged Bill Jenkins.[53]

The union could certainly take heart from its early efforts. In March 1986, BASF's top U.S. managers openly expressed their concerns about the OCAW's corporate campaign by calling a press conference in Washington, D.C., in which they asked for a congressional investigation of the union's tactics. "OCAW is engaged in an ugly, vicious campaign of fear and smear," charged James C. Burris, BASF's vice-president of human resources. Earlier that year, Richard Donaldson had walked into the offices of the AFL-CIO's Industrial Union Department (IUD) in Washington, D.C., and had succeeded in obtaining a copy of an IUD manual that advised unions on how to conduct such campaigns. Donaldson, an amiable man, had successfully posed as an interested student of union tactics in order to obtain the document. The company used the manual to claim that it was the victim of an "outrageous" new tactic that was being increasingly adopted by organized labor. BASF argued that the campaign violated the "essence and purpose" of the National Labor Relations Act because it distracted the union from trying to settle the dispute.[54]

In seeking to halt the campaign, however, the BASF managers themselves highlighted the union's success in broadening the basis of the dispute and bringing it before an international audience. The company claimed, for example, that a "local labor dispute in Geismar, Louisiana" had developed into "an international incident involving the U.S. Department of State, allegations of Nazi war crimes, and the Paris-based Organization for Economic Cooperation and Development, as well as claims of evading U.S. law and false alarms of a Bhopal-like disaster here in the United States." BASF backed up its complaints by filing charges with the NLRB that the campaign represented unfair labor practices. The union, claimed the company, had sought to "arouse public animosity, distrust, fear and suspicion of BASF by . . . asserting that BASF is a safety hazard to the community."[55]

In a September 1986 ruling, the General Counsel of the NLRB dismissed the company's charges, asserting that corporate campaigns were an accepted and legitimate weapon for unions to use. This decision was vital for the union and caused great consternation to BASF executives. Les Story admitted that he was "very disappointed" with the ruling, but he accepted that it would not be appealed further. "We're going to have to live with the corporate campaign," he said. Leonard and Miller were jubilant; they knew that their corporate campaign was affecting BASF, and they were now ready to take it to a new level.[56]

The Campaign Escalates

Throughout 1986, Leonard and Miller escalated the union's campaign against BASF. Leonard spent most of his time in Denver, working to create international pressure against the German chemical-maker, especially by cementing his links with the German Green Party. Miller, on the other hand, stayed in Louisiana, building contacts with local residents. Working tirelessly, he forged a bond between the union and local environmental groups that had not existed previously.

Upon arriving in Louisiana, Miller looked around for allies to support the union in its fight against BASF. Like the BASF plant, many chemical plants in Louisiana were located close to predominantly African-American communities. These residents had initially welcomed the plants into their communities, hoping that they would provide well-paying jobs, but by the 1980s many had become disillusioned. Companies insisted that local people lacked the qualifications to perform the skilled chemical jobs, turning instead to white workers who lived farther away. Many black residents felt that they were exposed to pollution without receiving any economic benefits in return. As such, they could not see pollution as a trade-off for well-paying jobs, as most BASF workers did. In the late 1970s and 1980s, residents of several communities consequently became increasingly outspoken against industry.[1]

In the early 1980s, for example, the residents of Alsen, a predominantly African-American community located to the north of Baton Rouge, led a high-profile fight against Rollins Environmental Services, which operated a large hazardous waste landfill that local people blamed for a wide variety of health problems. Rollins was a disposal center for major companies located along the chemical corridor, including BASF and Exxon, and it helped to ensure that Louisiana was home to one-third of America's hazardous waste landfill capacity. In 1981, Alsen residents brought a multimillion-dollar class-action lawsuit against Rollins. For more than six years, the case dragged through the courts before it was eventually resolved through an out-of-court settlement that only partially addressed residents' concerns. Nevertheless, the residents' activism has clearly pushed the company to improve its environmental performance.[2]

In Willow Springs, Calcasieu Parish, meanwhile, African-American farmers voiced their complaints against Browning Ferris Industries (BFI), claiming that the company's large landfill had contaminated their groundwater and had given them respiratory ailments, skin problems, blood disorders, and headaches. In 1980, local activists brought a legal action against BFI, but this case, too, dragged through the courts as the plaintiffs' lawyers labored to prove that the company was responsible for residents' health problems. In Reveilletown, an African-American community founded by freed slaves after the Civil War, residents formed Victims of a Toxic Environment United (VOTE United) to fight against the pollution from the nearby Georgia Gulf plant. In filing a 1984 class-action suit, they consciously disproved the common perception that "black folks aren't concerned about the environment." Their activism eventually pushed Georgia Gulf to pay to relocate the town's entire population in new houses farther away from the plant.[3]

Many residents were helped by Willie Fontenot, the head of the Citizens Access Unit in the Louisiana Justice Department's Land and Natural Resources Division. Fontenot owed his position to William J. Guste, Jr., Louisiana's attorney general between 1972 and 1992. A liberal Democratic millionaire, Guste was a member of a powerful New Orleans family that owned land and businesses in the Crescent City, including Antoine's Restaurant. Guste viewed his position as "the peoples' lawyer" and felt that his office needed to do more to help citizens tackle Louisiana's "innumerable" pollution problems. On a friend's recommendation, the attorney general hired Fontenot in 1979 in order to help citizens organize to tackle pollution problems.[4] Fontenot followed his mandate faithfully; crisscrossing the state in the early 1980s, he was successful in encouraging a considerable amount of opposition from citizens to industrial pollution. For many years, for example, he worked faithfully with the residents of Alsen, encouraging them to keep diaries that documented when major discharges from the plant occurred. In his exploration of the state's environmental movement, writer Jim Schwab called Fontenot "the grandfather of Louisiana environmentalism."[5]

Fontenot is a modest, genial man who shies away from such compliments. Nevertheless, his effectiveness in encouraging citizens to challenge corporations' environmental practices was stressed by Guste, who remembered that industry representatives frequently urged him to fire Fontenot: "They would come to me, delegations, and urge me to fire him for all the terrible things he was doing. . . . Their criticisms were just to get rid of him, because he was a fly in the ointment." With Guste's backing, however, Fontenot clung onto his position.[6]

Fontenot's success also reflected the scale of the environmental problems facing Louisiana. By the early 1980s, more than thirty years of largely unrestricted industrial growth had led to severe pollution problems. More toxic chemicals were released into Louisiana's environment than in any other state; between 1987 and 1989, for example, more than two billion pounds were emitted. Scientists increasingly linked this pollution to residents' health problems. The state was termed "a public-health catastrophe" by Dr. David Ozonoff, an environmental epidemiologist at Boston University's School of Public Health. A former physician at the Ochsner Clinic in New Orleans, meanwhile, claimed that the area between Baton Rouge and New Orleans was "like a massive human experiment conducted without the consent of the experimental subjects." By the mid-1980s, the state's high cancer rate in particular was increasingly being linked to industrial pollution.[7]

By the early 1980s, the press, a loyal supporter of industrial development in previous decades, was also beginning to raise concerns about the environmental consequences of the state's rapid industrialization. Threats to the quality of local seafood, a commodity valued by many state residents, produced some of the first concerns. In 1980, for example, a spill of more than twelve tons of toxic pentachlorophenol into the Mississippi River Gulf Outlet following a collision of two ships led to worries that shrimp in the river would be contaminated.[8] By the middle of the decade, environmental concerns had become much broader based than this. In 1985, both the *New Orleans Times-Picayune* and the *Baton Rouge Morning Advocate* published special editions that graphically highlighted the state's environmental problems. Calling Louisiana "the poisoned land," the *Times-Picayune* claimed that the oil and chemical industries had left an "ugly imprint" on the state. "The landscape is pockmarked with leaking waste pits, faulty industrial injection wells and a myriad of other sites where toxic debris is dumped haphazardly," it concluded. Both newspapers also expressed concern about the high incidence of cancer in the state.[9]

Located just outside the BASF plant gates, Geismar had originally been a racially mixed community, but as the plants moved into the area in the 1950s and 1960s, white residents began to leave. By the time of the lockout, the town was overwhelmingly African-American. "All the white people moved out from Geismar," recalled longtime resident Amos Favorite. "They moved out and went somewhere else and left us, left the black people here alone. So we just didn't have nowhere to go."[10]

The residents of Geismar had many of the same grievances as those who lived near chemical plants elsewhere along the Mississippi corridor. They complained of respiratory problems, particularly at night, and worried about contamination to their drinking water, which was drawn from private wells. Older people claimed that there were fewer fish to catch in the Mississippi River than there were before the plants arrived. Many residents also complained that their communities were suffering from an escalating cancer rate. Favorite, who moved to Geismar as a child in the 1920s, had seen nine members of his family die of cancer by the mid-1980s, and he blamed the plants. "I could see where the wrongness was," he recalled. "You could smell the stuff in the air and at night you'd be laying in the bed and you would breathe that stuff in the air you're taking in to live." Very few Geismar residents worked in the plants, however, and they lacked a close knowledge of what was taking place in them. Many felt powerless and alienated. "There was a lot of things they was doing wrong out there, and hell, the authorities didn't give a damn about us," asserted Favorite.[11]

When Richard Miller arrived in Louisiana, he called a meeting at the Geismar firehouse to hear residents' views about the environmental problems they felt were affecting their area. "When I got there, I looked around for allies," recalled Miller. "I thought, 'Who are the other victims?' The residents of Geismar were the first ones I saw." Miller was also encouraged to reach out to the residents by Willie Fontenot, who sensed that the two groups could form a powerful alliance together.[12] Many residents were too afraid to attend the meeting, but Amos Favorite went and told Miller about their complaints. Favorite was more outspoken than many, a trait that he linked to his service in World War II: "The army taught me that. I wasn't afraid of nobody."[13]

After the war, Favorite had participated in the civil rights movement, marching with Dr. Martin Luther King on the famous Selma to Montgomery protest of 1965. He was well aware that being an activist could jeopardize his own safety; in the late 1960s, he had even chased away the Ku Klux Klan from his house after his daughter became the first African-American to try to enroll in the nearest all-white high school. In his seventy-eighth year, he was able to laugh at the incident: "The Ku Klux Klan come in here. . . . I filled one of them's behind full of number six shot out there one night, through this bathroom window on this house here. . . . he was running down that highway. He said his ass was burning up with fire. He was running and hollering, and I pouring the oil on it." At the meeting convened by Miller, Favorite agreed to become the first president of a new environmental group called Ascension Parish Residents against Toxic Pollution (APRATP). It was the start of an enduring alliance between the Geismar residents and the locked-out workers.[14]

In the 1960s, Favorite had also led a class-action civil rights lawsuit to improve job opportunities for African-American workers at the Ormet Aluminum plant, where he had worked since the war. At Ormet, Favorite had been an active member of the United Steelworkers, and his background as a union worker helped him to forge his alliance with the OCAW. He had a good understanding of labor issues and expressed support for Local 4-620's members, calling the lockout "a mockery of justice and fair play" and terming the company's contract offer "shameful." The company, he argued, should "sit down with the union of our locked out citizens and negotiate in good faith with our union leaders." He formed a genuine bond with the locked-out workers and was still friends with many of them more than a decade later.[15]

Organized in early 1986, the APRATP claimed that its goal was to "improve the environment and leave a healthful quality of life for our children and grandchildren to enjoy."[16] An unincorporated neighborhood organization, the APRATP pressed for environmental protection laws in the Louisiana legislature. Its members were primarily African-American residents living in Geismar who were no longer willing to put up with the pollution caused by the plants. "We are sick and tired of seeing our families and friends dying of cancer or being diagnosed with cancer," asserted an APRATP statement. "Our water has chemical odors, and looks like milk coming out of our drinking wells. There is young people in our community who is dying of or has been diagnosed with cancer. These young people are inocent [sic] victims of the money Hungry polluters we have in our community. There has been an increase in still births in the area also. There is tons and tons of toxic chemicals released by these chemical companies in our community."[17] The group called for more air monitoring in

Geismar and for independent checking of their drinking water. They also spoke out against the fact that very few Geismar residents worked in the plants, asserting that local people were not receiving any economic benefits from the chemical industry.[18]

The alliance between the locked-out workers and the Geismar residents was mutually beneficial. As very few residents had traditionally been hired in the plants, community members often lacked a close knowledge of what occurred inside the gates. The locked-out workers supplied this, and the international union also provided the resources to carry out specialized environmental testing. During the course of the dispute, the union hired toxicologists and hydrologists to investigate pollution around the plant. "This union," noted Roger Arnold, "gave the citizens around these plants and all the weapon to help fight these plants." Amos Favorite agreed. "We're the ones that teamed up and we started raising hell," he recalled. ". . . We stuck in together, and then we start achieving something. . . . We was allies together when we first started fighting the pollution."[19]

For the union, meanwhile, the residents provided vocal allies in the community and allowed them to broaden the dispute beyond traditional labor issues. The contact with residents also made locked-out workers more aware of the needs of local people and the problems that they faced. "I passed through here every day," recalled Carey Hawkins. "I grew up passing through here on the way to the river to swim, fish, whatever. I never appreciated the day-to-day danger that they may have been in, through like a gas release, explosion, long-term exposure to chemicals, that there would be some long-term effects. I never really appreciated that until the lockout. Now, yeah, I'm much more understanding and much more sympathetic and much more conscious of the place I work. What do we do? Is it going to affect anybody just here or outside the fence also? Much more conscious."[20]

The APRATP soon secured one important change. Many Geismar residents were concerned about the fact that trucks carrying hazardous chemical waste passed through their community on a daily basis. In response to these concerns, Favorite wrote to the managers of all the chemical plants in Geismar to request that trucks traveling to and from the plants use Louisiana Highway 30, a broad road, rather than Highways 73 and 74, which were narrow roads lined with houses. In the past, there had been several minor accidents involving trucks traveling to and from the chemical plants on Highways 73 and 74, and residents worried that a more serious incident was inevitable. "Such an accident with its tragic consequences would tear the heart out of our community," noted Favorite.[21] The chemical plants were becoming very aware of the need for good public relations, and most of them quickly complied with Favorite's request, instructing their truck drivers to stay off Highways 73 and 74. The APRATP also secured permission for a "No Hazardous Materials" sign to be placed next to the interstate exit for Highways 73 and 74.[22]

Shortly after arriving in Louisiana, Miller also made contact with the Delta branch of the Sierra Club. The branch had been set up in 1969, a time when the number of grassroots environmental groups mushroomed across the United States. It was based in New Orleans and was led by Darryl Malek-Wiley, a white former chemical worker and union activist who was keen to get more involved in labor issues. In 1973, the national Sierra Club had worked with the OCAW during a strike and boycott against Shell Oil Company. The union struck the company to ensure the setting up of joint labor-management health and safety committees, a demand supported by the environmental organization. Malek-Wiley had been impressed with the alliance formed between the OCAW and the Sierra Club, and he was friends with several club leaders who had been involved with the boycott. When he heard about the dispute at Geismar, he felt that the company's position was "crazy" and was keen to work with the union to help end the lockout.[23]

Malek-Wiley admitted that his interest in labor issues set him apart from many members of his branch, who were largely middle-class whites who wanted to concentrate on bird-watching and wilderness issues. Some of these members indeed opposed his involvement with the union. "Environmentalists are white, middle-class," he asserted. "It was easier working with the guys in the local [union] than it was with fellow environmentalists. A lot of environmentalists told me, 'You're crazy, labor guys are not our friends, and why are you doing this?'" Malek-Wiley, however, felt that workers could be a powerful ally that could help environmentalists to tackle Louisiana's pollution problems.[24]

For much of 1986, Miller and Malek-Wiley worked together to produce pathbreaking reports that documented air and water pollution in the Geismar area. Willie Fontenot, an invaluable source of information about the chemical industry in Louisiana, helped the two men compile the reports, opening up his own files that documented pollution along the chemical corridor. The three activists also took advantage of Title III of the Superfund Amendments and Reauthorization Act of 1986, which required a broad range of industrial facilities to release an unprecedented amount of information about quantities of toxic chemicals released into the land, air, and water.[25] A variety of local environmental groups lent their support to the reports, including the APRATP and the Alliance against Waste and to Restore the Environment (AWARE), an Iberville Parish group. The first report concentrated on fifteen chemical plants in St. Gabriel, Geismar, and Plaquemine and showed that these plants discharged over 75,755,792 pounds of pollutants into the Mississippi River between October 1985 and October 1986. Another report, this one on air quality, showed that over a nine-month period, 196 million pounds of pollutants were released into the air by eighteen Geismar-area chemical plants. This amount equaled 46,000 pounds for each person living in the study area.[26] Darryl Malek-Wiley linked these releases to the area's high cancer rate: "If you have 18 plants then we're talking almost 200 million pounds, we're talking statewide, we're talking maybe a billion pounds of pollutants per year and they wonder why we have a high cancer rate. That might be something involved in it."[27]

The air report was also endorsed by LEAN, a new environmental group created out of a leadership development conference held at Louisiana State University in March 1986. The conference was organized by the Citizens Clearinghouse for Hazardous Wastes, a national environmental group. It marked the first-ever statewide conference on toxic pollution and public health, and it concentrated on the steps that citizens could take to tackle these problems.[28] LEAN was organized, as the group's founding charter put it, in order to "foster cooperation and communication among individuals and organizations to address the environmental problems of Louisiana." A federation of grassroots groups, LEAN was an attempt to create a unified statewide environmental movement.[29]

Among those who attended the conference was Marylee Orr, a local mother with a chronically ill child. Orr was very concerned about the effects of industrial pollution on residents' health. As she recalled, her activism grew out of these concerns rather than formal training. Aware that "there weren't very many women doing what I was doing," Orr pressed ahead because she felt "called" to become an environmental leader. In 1984, she started a group called Mothers against Air Pollution, lobbying plants in Baton Rouge to reduce pollution levels. Two years later, she volunteered to take part in the LSU conference, becoming a cochair of the embryonic LEAN. For two years, the group met in Orr's living room. A forceful, vibrant personality, Orr soon became executive director of LEAN, a position that she still held in the summer of 2000.[30]

Shortly after LEAN was organized, Miller became friends with Orr and Local 4-620 became a member group of LEAN. Ramona Stevens, the wife of a locked-out worker, was hired as a staffer by the fledgling group and was soon a close friend of Orr. Through Miller and Stevens, Orr met several locked-out workers and came to appreciate that workers were often the first to be exposed to toxic chemicals. Looking back, Orr felt that because of the lockout she had acquired a strong "commitment to workers' health and safety . . . that I don't think when I started this journey I was aware about." Realizing that workers often possessed detailed knowledge of companies' environmental practices and were a valuable potential ally, Orr encouraged worker participation in LEAN, both during and after the lockout.[31]

In the fall of 1986, the union joined with the Sierra Club, LEAN, and the APRATP to launch a successful campaign to protest the leniency of an air pollution penalty imposed on BASF by the state DEQ. Earlier in the year, the agency had found BASF guilty of five separate violations of the Louisiana Environmental Quality Act and the Louisiana Air Quality Regulations. These violations included a release of phosgene, the most dangerous chemical processed at the site. The company also failed to take action for fifty-five hours following a release of 16,000 pounds of toluene, a mutagen and embryotoxin. BASF also violated its state air emission permit for at least seven years; the company released 140.8 tons a year of toluene, exceeding the 4.4-ton-a-year limit set in its permit. Finally, BASF exceeded its legally permitted air pollution levels for nitrogen oxides, sulfur dioxide, and hydrochloric acid from a chemical waste incinerator.[32]

These violations occurred between April and June of 1986. During the summer of 1986, DEQ representatives met privately with BASF managers, who accepted responsibility for the violations and agreed to pay a fine of $66,700. The Sierra Club, the APRATP, and LEAN came together to file a joint appeal against this penalty. Their protest was based on a number of grounds. They argued that the DEQ should have allowed the public access to its meetings with BASF. As Amos Favorite put it, "We are tired of DEQ cutting sweetheart deals at our expense."[33] The environmental groups also asserted that the penalty was insufficient given the "economic benefits" that BASF gained from polluting. Relying on the methods used by the EPA for calculating penalties, the Sierra Club estimated that the DEQ fine captured only 2 percent of the economic benefits realized by BASF for violating air quality regulations. "Such a minuscule penalty," charged Malek-Wiley, "not only fails to deter future violations, but it sends an unfortunate message to industry—it pays to pollute."[34] Finally, the appeal pointed out that Attorney General William Guste had not given his concurrence to the fine. Guste had met with the Sierra Club in December 1986 and had decided, after hearing the organization's concerns, that the penalty was insufficient. The environmental groups argued that under the state environmental code any penalty had to have the concurrence of the state attorney general.[35]

Although the OCAW did not intervene openly in the case, Richard Miller was heavily involved behind the scenes. Miller contacted the EPA and was able to compile detailed data about the company's emissions for the Sierra Club. He also worked with Willie Fontenot, who was quickly becoming one of his most effective allies. Fontenot used his contacts with Guste to secure the attorney general's support for the environmental groups' appeal.[36]

BASF replied to this appeal by asserting that the citizens' groups should have requested a public hearing prior to the assessment of the penalty. It also argued that the DEQ could negotiate "compliance orders and/or administrative civil penalties" without obtaining the concurrence of the attorney general. The company rejected the assertion that the fine should be increased, insisting that it was "entirely appropriate."[37]

Although the litigation proceeded slowly, the First Circuit Court of Appeals eventually ruled that the government agency had failed to recapture the economic benefits realized by the polluter from noncompliance with air pollution regulations. This was a major breakthrough, marking the first time that a citizens' group was able to get the courts to increase a penalty because it failed to deter future noncompliance. In the past, in contrast, industry had often been able to secure court rulings that reduced its fines. The court also ordered the DEQ "to provide persons with a real interest an opportunity to petition for a hearing," and asserted that the attorney general should have concurred in any penalty assessment.[38]

While Miller was successfully building links with grassroots environmental groups in Louisiana, Richard Leonard reached out to environmentalists in West Germany as part of a broader effort to create pressure against BASF within Europe. In the mid-1980s, the Greens emerged as an important force in West German politics. Buoyed by public concern about radioactive fallout from the April 1986 explosion at a nuclear power station in Chernobyl, Ukraine, the party made big gains in the 1987 federal election, securing between 12 and 19 percent of the vote in many constituencies.[39] From his base in Denver, Leonard corresponded frequently with the leaders of the party and obtained helpful information from them, including the addresses of BASF shareholders.[40] Leonard also made several trips to West Germany, sharing a platform with Green Party leaders who criticized BASF's conduct. At one conference held in Bonn, Heinz Suhr, a member of the West German parliament, asked the company to "abandon its deeply anti-social and dangerous policy."[41]

In April 1986, Leonard helped organize a trip to Louisiana by a delegation of Green Party leaders. The Green delegation began their trip by visiting the union hall and meeting locked-out workers. After this, they demonstrated outside the plant gates with WSG members. The group, which included two members of the German parliament, tried to visit the plant, but BASF refused them entry. The Green leaders insisted that they had a duty to investigate the safety of the plant, especially in the wake of Bhopal. In response, BASF officials argued that the Germans were interfering in a local dispute and that it was futile to show them around the

10. Leaders from the German Green Party, led by parliament members Heinz Suhr and Willi Tatge, being denied entry to the BASF plant in Geismar. (Courtesy PACE International Union)

11. German Green Party leaders protesting outside the plant with members of the Women's Support Group. (Courtesy PACE International Union)

plant because they had already made up their mind to support the union. Executives dismissed the Greens as "far-out leftists" and argued that they had come to Louisiana to embarrass the company. They were strongly supported in this position by the LABI. "If it is the intention of the Green Party and the OCAW union to send signals around the world that Louisiana union members consort with radicals, then they have succeeded," stormed Ed Steimel, who protested about the visit in a telegram to the West German ambassador in Washington. OCAW leaders were pleased with the publicity they gained from this incident; several papers ran pictures of the German delegation being refused entry to the plant and detailed the union's safety allegations.[42]

In June 1986, Leonard made the first of four annual trips to Germany in order to address the company's main shareholders' meeting. He had bought some stock in BASF in order to give him the right to attend the meeting in Ludwigshafen, at which any shareholder was allowed to ask questions of the Vorstand (Board of Directors) and the Aufsichtsrat (Board of Supervisors). While in Europe, Leonard aimed to spread news of the lockout and exploit possible divisions between BASF's European and North American managers. BASF management in North America consistently asserted that they had complete control over labor relations, but Leonard was keen to see if this was really true. He prepared extensively for the trip, brushing up on his German, since the shareholders had to be addressed in their native language. As speakers were not subjected to a time limit, Leonard presented a detailed outline of the Geismar lockout in front of 2,500 worker-shareholders, most of whom lived and worked at the main plant in Ludwigshafen. He called the lockout "an aggression designed to obliterate the union" and called on the Vorstand to "exercise its responsibility in this conflict and bring it to a humane and speedy conclusion." He also submitted a resolution seeking a payment of $25 million to the locked-out workers in order to compensate them for their lost wages.[43]

BASF's response was given by chairman Hans Albers. In a terse statement, he denied that the Geismar plant was unsafe and asserted that German management would not intervene in the dispute. "Labor relations," he asserted, "is a responsibility of the local management and BASF Corporation in the United States. BASF will not be blackmailed by anyone, anywhere." Leonard's proposal was rejected.[44]

In Geismar, local managers asserted that Leonard's visit had achieved little, and they posted Albers' remarks in the plant. Leonard, however, was keen to try and build support for the lockout in Europe, and he was pleased that he had been given the opportunity to present the union's case to a new audience. The OCAW activist was able to cement links with a few BASF workers in Germany, and they began to send information back to the United States about the German company. One of these workers, Bernhard Doenig, visited Geismar with the German Greens.[45]

Leonard found, however, that his links with the German Green Party further alienated the leadership of I.G. Chemie. Like most German unions, the chemical workers disliked the Green Party, frequently attacking it on the grounds that its programs threatened "economic insecurity."[46] Afraid that the OCAW's support of the Greens would alienate the German union further, IUD strategists urged the American union to cut its ties with the German environmentalists. In December 1986, the IUD's Joe Uehlein argued that if I. G. Chemie could be brought on board, they would be more influential in persuading BASF to settle the dispute. "On the international level, I remain convinced that maintaining good relationships with I. G. Chemie is critical," he wrote Leonard. "After all, they have the influence with BASF; power, if you will. The Greens do not." Pursuing this strategy, Uehlein even persuaded AFL-CIO president Lane

Kirkland to write I. G. Chemie and request "more substantial assistance" from it in settling the dispute.[47]

Leonard was unmoved. By the end of 1986, he had lost all patience with I. G. Chemie and remained convinced that the alliance with environmentalists was the most effective way of settling the dispute. The OCAW had made "every possible attempt" to secure the goodwill of I. G. Chemie, he argued, but the German union was simply unwilling to help them. For Leonard, the German union's refusal to support the OCAW's accusations about the safety of the plant was clearly the last straw. "The I. G. Chemie has denounced our conclusions about the safety conditions at Geismar," he wrote. "These are not just simple misunderstandings, but are part of a deliberate pattern to trash us out."[48]

In seeking to work with I. G. Chemie, OCAW leaders had hoped to capitalize on the differences between the company's stable relationship with unions in Europe and its harsh treatment of unionized workers in the United States. In fact, these differences made it difficult for the Germans to understand events in Geismar, as OCAW leaders repeatedly complained that I. G. Chemie leaders were refusing to believe that BASF was trying to weaken them. After a detailed analysis, Leonard concluded that I. G. Chemie's experience of stable collective bargaining made it impossible for them to comprehend the lockout. "This organization has not had any kind of a fight since the '50s," he noted. "I don't think they would know how to fight or defend themselves if it came down to it. They don't have any comprehension about what we are up against." Overall, I. G. Chemie leaders did express support for the Geismar workers, donating $10,000 to Local 4-620 in March 1986, but they failed to act in the decisive or wholehearted way that OCAW leaders had hoped for. The lockout was only covered occasionally in the I. G. Chemie newspaper, and even then it was reported in a dry and factual way.[49]

In his efforts to exert international pressure against BASF, Leonard also sought to gain the support of unions across the globe, particularly those that represented chemical workers. He corresponded diligently with chemical workers in other countries and had some success in securing their backing for the OCAW's struggle. The General Chemical Workers Union in Malta, for example, sent an expression of support, as did chemical workers in Brazil who were also fighting for union recognition at a BASF plant. Several overseas unions also donated money to Local 4-620. Fuller cooperation was often held back by the language barrier, as well as the fact that these unions were often consumed in their own struggles.[50]

The OCAW also worked with the International Federation of Chemical, Energy, and General Workers' Unions (ICEF), a worldwide organization with more than two hundred affiliated trade unions from seventy-five countries. The ICEF represented more than six million workers in process industries and acted as a service center for its affiliates, providing research facilities on multinational companies. Throughout the lockout, ICEF offices in Geneva and Brussels provided the OCAW with information on BASF that helped the American union devise its corporate campaign strategies. The international federation's Geneva offices, for example, helped Leonard collect information on the Bhopal tragedy.[51]

Conducting extensive research into BASF's business operations across the globe, Leonard found that BASF was supplying Hitachi IBM-compatible computer systems to Persetel, a subsidiary that in turn sold them in South Africa. In November 1986, he set about publicizing the company's trade links with apartheid South Africa in an effort to create negative publicity for the German company. Working through a consultant, the OCAW commissioned an open letter that called on BASF and Hitachi to end all computer sales in South Africa. The letter expressed particular concern in light of an August 1986 New York Times article that had revealed that Hitachi computers were being sold to the South African police, despite the fact that West German export regulations barred such sales. In response, both BASF and Hitachi asserted that they complied with all relevant restrictions on computer sales to South Africa and would continue to supply computers for sale there.[52]

The union's South Africa campaign drew support from the antiapartheid movement in the United States, Germany, and Japan. The open letter was also signed by the leaders of several other unions, including Mineworkers' president Richard L. Trumka and Steelworkers' head Lynn Williams. Charging that BASF was "becoming a collaborator in the brutal system of apartheid in South Africa," Democratic presidential candidate Jesse Jackson also lent his support. The open letter was well publicized, both in the United States and Germany. The leading German weekly Der Spiegel, for example, noted that BASF's trade in South Africa was coming under "heavy suspicion" and questioned whether the company was "the ugly German."[53]

Although BASF's sales continued, the union leaders sensed that their campaign had made the company "uptight." In fact, company files indi-

12. Jesse Jackson expressing his support for the locked-out workers. To his immediate left are Richard Miller and locked-out worker Leslie Vann. Jackson is also flanked by Women's Support Group members, state representatives, and labor leaders. (Courtesy PACE International Union)

cate that the union's South Africa campaign did worry BASF a great deal. The company's PR consultants highlighted this in one letter: "This is a very sensitive issue in the United States and could be used very effectively against BASF. Few politicians are willing to resist the antiapartheid effort, even when it is at its most unreasonable extremes." Jackson's support for the campaign also concerned BASF; they recognized his power and ability to generate media interest, describing him as "a convenient and potent ally for OCAW." The company's sensitivity suggests that the union could have profitably devoted more resources to the South Africa campaign, but instead environmental work gradually pushed the South African campaign into the background.[54]

At a Baton Rouge press conference held a few weeks later, Jesse Jackson also gave his broader support to the locked-out workers. In a detailed condemnation of BASF, he likened the lockout to racial discrimination, claiming that both were "immoral as well as illegal." "Labor laws were passed by Congress to protect workers from employer discrimination, just as civil rights laws were passed to ensure that the constitutional rights of minorities were protected," he said. "Congress never intended to legalize the offensive lockout." Flanked by locked-out workers, union leaders, and black state legislators, Jackson also claimed that the plant posed an environmental danger because it was operated by poorly trained temporary workers.[55]

Jackson's support was a morale boost to the locked-out workers, who often felt that state politicians had not supported them enough. Over two hundred workers signed a letter of thanks to the civil rights leader, while others wrote him personally. "Your speaking out against this company was morally uplifting and motivating to our Union membership," noted one.[56] OCAW leaders were understandably delighted to have received backing from such a prominent politician. In September 1988, moreover, Jackson returned to Louisiana to drum up support for the Democratic Party in the forthcoming presidential election. On a visit to Baton Rouge, the veteran civil rights leader met with a delegation of locked-out workers and environmentalists and said that greater care must be taken to preserve the environment.[57]

BASF's concern about Jackson's support for the union highlighted the fact that the union's campaign was exerting some leverage. While usually seeking to downplay the impact of the union's efforts, one unnamed BASF executive commented as early as January 1986 that the union's campaign was "very sophisticated" and representative of "a whole new ball game in labor relations." By switching the focus to environmental issues, the OCAW had seized the initiative from BASF. The union's alliance with local environmental groups drew particular comment from former managers. Richard Donaldson thought the union's bond with local residents was "clever," while Les Story admitted that the move had "surprised" him.[58] Company records indicate the close interest that BASF took in the union's campaign. Every move by Leonard and Miller was closely watched and recorded, with Les Story being given frequent written briefings by both Richard Donaldson and public relations consultants. In November 1986, for example, PR consultant Rafael Bermudez warned local managers of Richard Miller's return to the area after a trip away and predicted that the young consultant was likely to play an increasingly important role in the union's campaign. "It is my feeling that OCAW is gearing up for a new round of harassment tactics against BASF," he added.[59]

Although clearly concerned about the campaign, BASF officials showed few signs of willingness to settle the dispute. The company's public relations consultants urged managers to hold firm, arguing that the union had not been able to recruit several key figures to its side. They stressed that BASF was still supported by local business leaders and that influential political figures in the state had remained neutral. "No major political figure has appeared on television or been quoted in the newspaper as supporting OCAW," noted one document. Governor Edwards, they pointed out, had been given "plenty of opportunities to blast the company" but instead had gone "out of his way to remain neutral." The PR consultants also assured BASF that the union's environmental campaign was unlikely to work because the general public would see that the union was simply trying to use the issue to try and gain leverage at the bargaining table. They stressed that the *Baton Rouge Morning Advocate* was still supportive of the company's position, adding that several BASF executives had good contacts with staff at the paper. They also argued that allowing workers back into the plant would enable them to carry on the corporate campaign inside the facility and could jeopardize production.[60]

Apparently heeding this advice, the company failed to give ground in negotiations. For much of the dispute, indeed, talks floundered because of fundamental differences between the company and the union, particularly over the subcontracting of maintenance. BASF insisted that it had to subcontract the maintenance jobs, as well as modifying the contract's seniority provisions and holding back wages, in order to make the Geismar site more competitive. "The longterm growth and stability of the Geismar site is the central issue in this labor dispute," wrote Story. The union flatly refused to accept the subcontracting of maintenance and argued that every union member who was locked out should be given the chance to return to the plant. Union leaders felt that it was their duty to fight for their members' jobs in this way. Richard Leonard summed up the union's position in a letter to Methodist bishop George Ogle: "Our bottom line is that as long as there are jobs available, any union member who wants to go back, should be allowed to do so. . . . George, can we really do any less than this and still call ourselves a Union? Can we represent our members and at the same time ask them to resign? Can we really do anything less than take an 'absolutist' position on the question of existence?" Local union members remained strongly committed to fighting until all their members got the chance to go back to work. "That was our intent from the beginning," recalled Carey Hawkins, "that we all wanted to go back to work and we couldn't pick or choose a job, we just wanted to get our people back to work."[61]

The union did try to offer concessions in order to retain the jobs of its maintenance workers, but BASF was in no mood to compromise. In August 1986, for example, BASF refused an offer that all Local 4-620 members would take a $2-an-hour pay cut if they were allowed back in the plant. The company rejected the offer because it was contingent on the company's keeping the OCAW maintenance workers, and BASF officials insisted that this would cost them more than they would save through the pay cut. The company also argued that the pay cut would cause greater turnover and training costs, as it would cause workers to leave for higher wages elsewhere. BASF officials furthermore insisted that they wanted to employ contract maintenance workers because of their "flexibility" and willingness to work long hours, in addition to the direct economic benefits. Following this refusal, negotiations once again reached a deadlock.[62]

Assisted by its public relations consultants, BASF also hit back at the union's campaign with a series of assertive press releases. Insisting that they were not antiunion, executives stressed that they had settled contracts with unions in other locations. In particular, the company pointed to an agreement it had successfully concluded with an OCAW local at its Rensselaer, New York, plant in September 1984. Union leaders, however, asserted that contracts had been settled only when the unions had "rolled over" and agreed to the company's concessionary demands.[63] In Rensselaer, BASF itself noted that the 1984 contract required increased worker contributions to the health plan and included "an extremely strong Managements' Rights provision." BASF's top managers felt that the Geismar union should have followed suit and shown more understanding for the company's competitive position, yet they never fully comprehended the union's feeling that this would have undermined its status as the workers' representative.[64]

As the dispute dragged on, many locked-out workers faced mounting financial problems, with many losing their homes and other possessions as a result. BASF was fully aware of this; a survey conducted by the company in February 1986 shows that the company was documenting the number of workers who lost houses and cars as a result of the lockout. A single mother with three children, Gladys Harvey lacked the family support that many workers had. Harvey vividly recalled the "tough times" that she went through during the lengthy lockout. She made many personal sacrifices to try and provide for her children, even selling her wedding ring and other personal jewelry so that she could buy a flute for her daughter to play at school. Unable to find another job in Louisiana, Harvey ended up losing her house and filing for bankruptcy. After this, she returned to her native New England to work in a paper mill for the rest of the lockout.

The experience had permanently affected her, turning her into a diligent saver so that she would not be caught in the same position again.[65]

In August 1986, the union instituted an "Adopt-a-Family" program to help locked-out workers who had been hardest hit by the dispute. The program was initiated by the OCAW's international executive board and was modeled on a similar effort launched by the UFCW during the Hormel strike. Following an extensive letter-writing campaign, twenty-seven OCAW locals agreed to make regular monthly contributions to the program, which was administered by the WSG. A board composed of WSG members assessed the financial situation of the locked-out workers' families and distributed funds accordingly. Applications were considered on an anonymous basis to prevent possible partiality, and individual families were sponsored by contributing local unions. Both locked-out workers and WSG members traveled around the country raising money for the program. While on their travels, the Geismar delegation visited other workers involved in labor disputes who were also fighting to resist concessionary demands. The Adopt-a-Family program certainly helped provide for cash-strapped families, as well as providing a clear sense of purpose to those who administered it.[66]

Throughout 1986, union leaders were confident that they were exerting some leverage against BASF, but they also accepted that they needed to do more in order to force a settlement. Leonard himself was convinced that the campaign was working, but he also credited the company with a "a fairly good showing," adding that it had acted "as we would expect from a formidable adversary." Aware that the campaign was going to be a long haul, he recommended that the union had to go on the "offensive" to increase pressure on the chemical giant. In 1987, he and Miller launched a string of new initiatives against BASF that finally broke the deadlock.[67]

6

Breakthrough

As 1987 began, Richard Leonard was optimistic. Aware that the international union remained fully committed to the BASF campaign, he privately predicted that he could secure "a successful conclusion to the lockout in the not too distant future." Later that year, the deadlock was indeed broken when BASF agreed to call all the operators back into the plant. The breakthrough came after the union had secured increasing support from environmentalists and other community leaders for its position. BASF became increasingly uncomfortable as the profile of the campaign continued to rise and called back the operators in an attempt to defuse the union's escalating campaign.[1]

Striving to find new ways of exerting pressure on BASF, Leonard, in the spring of 1987, came up with the idea of conducting a demonstration at the company's North American headquarters in Parsippany, New Jersey. Through the spring and early summer of 1987, he lobbied intensively for the event, soliciting endorsements from a wide variety of environmental groups, including Greenpeace, the Sierra Club, the National Campaign against Toxics, and the New Jersey Environmental Federation. Some churches, including the Baptists, also endorsed the protest, as did several other unions. Drawing on Greenpeace's tactics of civil disobedience, Leonard planned to take protesters to the company's headquarters and block all access points to the building. This "lock-in" of BASF staff was designed to publicize the way that unions felt that companies were increasingly using offensive lockouts against workers.[2]

On the afternoon of June 3, 1987, more than 350 protesters blocked the driveways of BASF's headquarters, preventing staff from leaving work. They chanted "Locked out—locked in, now you know the shape we're in," and displayed banners that read "Danger BASF Toxic," "Break the BASF Lockout," and "BASF Get Out of South Africa." A few protesters also staged a "die-in," gasping and dropping on the ground to highlight

13. Protesters blocking the road to the Parsippany offices during the "lock-in."
(Courtesy PACE International Union)

the hazards posed by the BASF plant. A series of speakers charged BASF
with creating conditions similar to those that caused the Bhopal disaster. For over ninety minutes, the back driveway of the headquarters was
chained shut while the front driveway was barricaded with a core of eighteen demonstrators. The Parsippany police had plenty of warning about
the demonstration and peacefully arrested the eighteen participants who
were blocking the front entrance. Among those removed from the scene
were OCAW vice-president Robert Wages and New Jersey Industrial
Union president Archer Cole.[3]

Although the OCAW secured the services of former U.S. attorney
general Ramsey Clark to represent those who were arrested, their municipal court trespassing convictions were upheld by New Jersey's Superior
Court. Clark argued before the court that the Geismar plant was unsafe
and that the protesters had a "right and duty" to bring public attention to
the situation. The veteran lawyer had been attorney general during many
of the civil rights crises of the 1960s and he tried to justify the protesters'
civil disobedience by comparing it to the tactics of these earlier demonstrators. Judge Reginald Stanton, however, was unmoved. Rejecting Clark's
arguments, he criticized the protesters' failure to obey orders from the
police to stop blocking the driveway to the BASF building.[4]

14. OCAW vice-president Robert Wages, being led away after his arrest during the "lock-in." (Courtesy PACE International Union)

For Leonard, the lock-in represented the culmination of his efforts to build alliances outside the labor movement and broaden the base of the BASF struggle. Pleased with the press coverage that the event had generated, he celebrated it as "an important step in the creation of a national coalition in support of the locked out BASF workers and opposed to the use of lockouts in general."[5]

The company had also planned well in advance of the lock-in. All Parsippany employees were mailed factsheets that outlined the company's case in the dispute. In them, BASF denied that safety had been compromised at Geismar and argued that the union should return to the negotiating table rather than protesting in other states. Many staff were allowed to go home before the protesters arrived, reducing the disruption caused by the protest.[6] Having made these preparations, BASF officials, not surprisingly, were keen to downplay the importance of the lock-in. Les Story claimed that it had "little impact" on the company in Geismar. He acknowledged, however, that the company disliked the publicity that the protest generated, adding that "nobody likes to have picketers show up in front of the building."[7]

Leonard tried several other strategies to exert pressure against BASF, including an attempted boycott of the company's products. A boycott had proved effective during the J. P. Stevens campaign, but against BASF the tactic was more problematic. By the early 1980s, most of chemicals made in the company's U.S. plants were used as components in other products and were not sold directly to the consumer. The company, for example, was the largest supplier of private-label antifreeze in the North America; in 1984, over 25 percent of all new cars sold in the United States had BASF antifreeze in them. In addition, the seats, armrests, dashboards, and soft facia of many new vehicles were made from chemicals produced at Geismar. The company did make one prominent retail product at Geismar—the broadleaf weed killer Basagran. The Basagran plant closed down

15. A protester being arrested during the lock-in. (Courtesy PACE International Union)

in 1987, however, leading to a major dispute between the company and the union about how to clean up the contaminated site.[8]

In July 1987, Leonard admitted that he had tried to launch a boycott of BASF products but had been unable to identify their consumer markets. As a result, he termed the boycott "really only a minor aspect" of the campaign. Several BASF executives asserted that the boycott efforts were ineffective. The company had no consumer products carrying its name, other than one herbicide and recording tape, which together constituted a very small part of their North American business. There were also no stockholders in the U.S. company other than the parent company. It was very difficult for the union to build support among BASF customers, as most of them were other large companies. "Almost all of our sales were to large Fortune 500 type organizations who had no brief in favor of organized labor and no reason to try to pressure us," asserted Henry Kramer.[9]

The union's boycott activities were not completely ineffective, however. In particular, executives clearly disliked the union's efforts to communicate directly with its customers. After compiling lengthy lists of their addresses, Leonard began to write BASF's customers in an attempt to inform them about the dispute, hoping that they might then pressure BASF to settle. These efforts annoyed the company. In October 1987, for example, BASF sought damages from OCAW leaders and attempted to enjoin the union from communicating with its customers, stating that its "reputation in the marketplace clearly has been damaged." These actions were eventually dropped following a November 1987 consent decree between the two parties. Shortly after this, however, the union mailed a letter to four thousand chemical industry executives. In it, the union claimed that the company's "anti-union" approach raised "serious questions as to BASF's judgement and ability to manage." Following this letter, BASF charged that the OCAW's leaders had violated the decree and petitioned for damages and sanctions. The union responded by defending its actions, claiming that the company was trying to prohibit its right to free speech as protected under the First Amendment.[10]

Leonard also worked with the AFL-CIO and was successful in getting BASF placed on the union federation's annual "dishonor roll." Started in 1983, the roll was an annual listing of companies that organized labor claimed were "notoriously anti-union." By placing a company on the list, the AFL-CIO hoped to encourage public scorn and thereby pressure that company to take a softer line with organized labor. Many union members, in particular, used the roll to avoid buying products made by companies

16. A delegation of union supporters in West Germany. Amos Favorite is on the far right, Richard Miller on the far left. The delegation also included Willie Fontenot (fourth from the right), Darryl Malek-Wiley (fourth from the left), and Richard Leonard (second left). (Courtesy PACE International Union)

that were listed. As with the boycott, BASF claimed publicly that the dishonor roll had "virtually no impact" on BASF business, largely because the company sold few products directly to the consumer. Again, however, company executives disliked the way that the list publicized the union's allegations about the company in the national press.[11]

In June of 1987, Leonard was part of a delegation to West Germany that included Richard Miller, Darryl Malek-Wiley, Amos Favorite, and Willie Fontenot. The group attended the company's shareholders' meeting, with Leonard again addressing the gathering. He tried to press the Germans to become involved in the lockout, but chairman Hans Albers again argued that the dispute was "a local matter for which our American management is exclusively responsible."[12] Leonard did, however, secure some press coverage in Germany, and he also again enlisted the support of the German Green Party, which issued a press release that declared the party's backing for the Geismar workers. Stressing the importance of unions and environmental groups working together, Willi Hoss, a spokesperson for the German Greens, offered the "unqualified support" of his party to the locked-out workers. He also praised the OCAW as being "one step ahead" of I.G. Chemie; the German union, he noted, "have so far been unable to decide to concretely support the fight of their American colleagues against the German BASF."[13]

Back in Louisiana, Richard Miller also continued to work on cementing the union's alliances with environmental and church groups. Throughout most of 1987, the OCAW strategist joined forces with the APRATP in order to fight a new hazardous waste incinerator that BASF wanted to operate at the Geismar site. The collaboration reflected Miller's efforts to use the regulatory process to disrupt production at the Geismar site and thereby exert economic pressure on the company. Citing the company's record of accidental releases over the previous two years, the APRATP argued that BASF could not be relied upon to operate the incinerator safely.[14] Locked-out worker Darryl Stevens worked as a research analyst for the APRATP and produced a detailed report which argued that the Geismar area was already too polluted. "Geismar is simmering in a gumbo of toxic and carcinogenic substances released from these companies," he asserted.[15]

In response, BASF argued that the incinerator would handle only non-hazardous waste and that it would reduce the heavy truck traffic that was required to dispose waste off-site.[16] The company also insisted that the incinerator was safe and would "have no adverse effects on the health and safety of the Geismar community."[17]

After lobbying the DEQ, the APRATP secured a public hearing of the BASF proposal. In September 1987, the hearing was held at the agency's headquarters and drew a crowd of nearly five hundred people, including supporters of both sides. After hearing the evidence, the APRATP's request for an adjudicative, or trial-type, hearing was granted by the DEQ.[18] The decision annoyed BASF; its representative asserted that delaying the permit would "have an adverse economic impact on our operations."[19] Working together, the union and the APRATP had clearly slowed BASF's permit; the company had decided to build the incinerator in November 1985, yet it still did not have the permit in April 1988.[20]

Miller also caused other problems for BASF. In the spring of 1987, the company had closed its Basagran herbicide plant. Digging around in state records, Miller noticed that BASF had failed to notify the state of its closure and was continuing to receive tax exemptions for the plant. Since these exemptions were provided as an incentive to build plants and create jobs, the law required them to be taken away when plants were closed. In September, Miller notified the Louisiana Board of Commerce and Industry that BASF was in violation of state regulations regarding ten-year industrial tax exemptions. After an investigation, the Board of Commerce and Industry agreed with the union's contention and canceled nearly $75 million in tax exemptions as a result. BASF's tax payments thus increased by over $508,181 in 1988. The episode revealed that parishes had no ongoing audit for compliance with regulations for industrial tax exemptions and could be missing out on substantial amounts of revenue as a result. The move was a public relations victory for the union and created some embarrassing headlines for BASF.[21]

Miller's work with the Sierra Club to document air and water emissions from the local chemical plants was to have far-reaching consequences. While compiling the report on air pollution, Miller and Malek-Wiley talked to several women in the local community of St. Gabriel who felt that there was a high miscarriage rate in their town. Local pharmacist Kay Gaudet had compiled a list of sixty-three women in the tiny community who had suffered miscarriages. Her figures meant that one of every three pregnancies there since 1983 had ended in fetal death, a miscarriage rate that was more than twice the state average. The OCAW joined with LEAN, the Sierra Club, the APRATP, and several women from St. Gabriel to spotlight their findings at a press conference held in the State Capitol.[22]

In the summer of 1987, the miscarriage story gained national attention when *USA Today* ran an article outlining the St. Gabriel women's concerns about the number of miscarriages that were occurring in their community. Three months later, the *Washington Post* picked up on the story and ran a detailed piece about the issue. Although the company suspected the union's hand behind the allegations, Gaudet had worked independently of Miller, although the two knew each other and Miller certainly encouraged her to pursue her work. He also supplied a package of information to *USA Today* that the paper used as the basis of its coverage.[23]

Miller recognized that the miscarriage story provided the union with a platform for publicizing the BASF dispute and the environmental issues that it raised. In the *Washington Post* article, for example, journalists David Maraniss and Michael Weisskopf interviewed the OCAW strategist and outlined the union's environmental allegations. They mentioned the "Bhopal on the Bayou" billboard and detailed the union's claims that pollution from the plants was adversely affecting residents' health. The media attention threw industry groups onto the defensive, and they struggled to respond effectively. Maraniss and Weisskopf called attention to an interview that Louisiana Chemical Association president Fred Loy had given with the student newspaper at Louisiana State University. In it, Loy, an outspoken defender of industry, had brushed off the miscarriage allegations: "They say the chemical plants are causing the miscarriages, but they have no proof. I could say that they screw too much and that's the cause of the miscarriages. But then I would have no way to prove that." Shortly after this quotation appeared in the national press, Loy was replaced as head of the LCA, and it was widely rumored that his response had cost him his job. His replacement, Dan Borne, was an accomplished public relations expert with a radio voice. Under Borne, the LCA became more conciliatory and less gaffe-prone.[24]

The press attention certainly helped the environmental movement in Louisiana. LEAN leader Marylee Orr, for example, wrote the *Washington Post* to thank it for its article. "I wish that words could express how beneficial that article has been in helping our organization," she noted.[25] Political leaders were pushed to react to mounting public concern. After both Democratic Congressman Buddy Roemer and state attorney general William J. Guste expressed their anxieties about the miscarriage rate, Edwin Edwards commissioned a state study that was carried out by Tulane University's School of Public Health and Tropical Medicine on behalf of the Louisiana Department of Health and Hospitals (DHH).[26] This study, which was supported by public funds, concentrated on St. Gabriel women between the ages of eighteen and fifty who conceived between April 1, 1982, and April 1, 1987, and had delivered between May 1, 1982, and December 31, 1987. It concluded that the area's miscarriage rate of 15.7 percent was "not statistically significantly higher than expected" and asserted that further study of the issue was not warranted.[27]

Environmental activists were not convinced by these conclusions. "They designed the study so that it would come back and say there were no problems," charged Darryl Malek-Wiley. Malek-Wiley was convinced that the miscarriage rate in St. Gabriel was higher than normal. He argued that the rate rose when there were specific accidental toxic chemical releases from plants near St. Gabriel but that the effect of these increases could be missed in a broad-based study. Union leaders and area residents were also skeptical of the report. They charged that the DHH had "hand-picked" community representatives and had excluded local residents from

17. The "Cancer Alley" billboard. (Courtesy PACE International Union)

the committee to establish the "protocol" for the St. Gabriel study. These concerns were also shared by new DEQ secretary Paul Templet.[28]

The miscarriage controversy remained highly contentious long after the Tulane study, although the study's publication did help the issue to drop out of the headlines. Kay Gaudet, the women who first raised her concerns about the miscarriage rate, has also found it difficult to continue her activism. Some chemical companies reportedly told their workers not to use Gaudet's pharmacy, driving her out of business. By the early 1990s, Gaudet was working in a chain pharmacy in Baton Rouge.[29]

Miller also realized from an early stage that BASF was sensitive to public pressure about the relationship between pollution and fatal diseases, particularly cancer. In 1985, Louisiana led the nation in lung cancer deaths and was ranked sixth in overall cancer deaths. A growing number of citizens and environmentalists linked the cancer rate to industrial pollution, a link denied by industry. Many of the chemicals produced in the chemical corridor were, however, suspected or confirmed carcinogens. Working through detailed medical records, Miller himself had found that a cluster of BASF workers who had handled the carcinogen formaldehyde had died of similar cancers. To replace the "Bhopal on the Bayou" billboard, the OCAW strategist came up with the idea of erecting a new sign to publicize the possible links between industrial pollution and cancer to passing drivers, as well as informing them about the lockout. In the summer of 1987, a new billboard asked, "Is BASF the Gateway to Cancer

Alley?: Residents and Locked Out BASF Workers Demand Industrial Accountability." Rather than a chemical corridor, the union argued that the communities between Baton Rouge and New Orleans made up a "cancer alley."[30]

The union had certainly succeeded in raising a topic that generated enormous public concern. Since the 1920s, cancer has been the second most common cause of death in the United States, and in 1985 the many forms of cancer accounted for the death of 462,000 Americans. Cancer, moreover, has always evoked tremendous popular fears about its deadliness. "Cancerphobia," notes historian James T. Patterson, "is deeply rooted in American culture."[31] From the 1960s onwards, increasing numbers of Americans began to blame industrial pollution for cancer. Many were influenced by Rachel Carson's *Silent Spring*, which concentrated on the carcinogenic effects of pesticides and herbicides. Some of the chemicals discussed by Carson were produced in Louisiana. The Vulcan plant located next to BASF, for example, emitted hexachlorobenzene, which Carson had linked to the fatal disease. Public opinion was also influenced by events such as the accident at the Three Mile Island nuclear power plant in Pennsylvania and the discovery of toxic chemicals dumped in the Love Canal in Niagara Falls, New York. Media coverage increasingly linked the chemical industry to cancer. In 1976, for example, a CBS special report titled "The Politics of Cancer" denounced the chemical industry.[32]

In Louisiana, the links between the petrochemical industry and cancer were hotly debated at the time of the lockout. In the summer of 1985, for example, both the *New Orleans Times-Picayune* and the *Baton Rouge Morning Advocate* ran special reports that explored the issue. The *Times-Picayune* cited a recent study by the Council on Economic Priorities, a nonprofit New York–based environmental research agency, which had found that cancer rates in areas of the state with heavy concentrations of petrochemical toxic waste were 68 percent higher than the national average. Industry, however, repeatedly asserted that lifestyle factors were the main reason for these rates.[33]

BASF managers dismissed the union's claims in a similar fashion. "The incidence of cancer in the state," recalled Les Story, "was primarily to do with the diet and the fact that almost everybody down there smoked heavily, including our adversary Mr. Rousselle." Some industry figures, however, admitted privately that they were concerned about the state's cancer rate and felt that industrial pollution was at least partly responsible for it. Former LABI president Ed Steimel remembered the billboard well: "'Cancer Alley' was what this area was called, and they [the union] were considered to be the principal cause of that, but a lot of that was hype, but it may have been. I don't think there's any doubt that the poor environmental standards that were held here contributed to cancer."[34]

Although some BASF managers are still loath to admit it, the union's "Cancer Alley" billboard did have a lasting impact on public opinion. More than a decade after the dispute, the billboard remained a lasting memory of the lockout for many state residents. "This many years later that was so effective people still talk about that billboard," reflected Marylee Orr. Orr felt that the billboard helped to break the lockout by creating public pressure against BASF. "It wasn't just workers and union people that were coming into them, it was community people, communities who were concerned. It was the general public who now saw the billboards, who read the articles. . . . I think it reached out to a broader group of people, and when you do that I think it's a lot harder for them to deny some of the problems they have." In the fall of 2000, Richard Donaldson credited the union for bringing to the fore the "Cancer Alley" debate that was still being conducted between industry, environmentalists, and concerned citizens. At the time, however, BASF managers were very annoyed with the billboard, claiming that it was misleading the public and damaging the state's economy and image. Both the "Bhopal" and "Cancer Alley" billboards, complained BASF executive Bill Moran, were "scurrilous and offensive."[35]

Industry was also annoyed when Oprah Winfrey traveled to Louisiana to tape an edition of her show that concentrated on residents' complaints that industrial pollution was responsible for cancer and miscarriages. One industry journal charged that the show was one-sided and claimed that the audience had "the collective cognitive powers of an equal weight of pea gravel." Winfrey herself was dismissed as "the daytime talk show host with the heart of gold and the brain of brisket." Another observer sympathetic to industry claimed that the show "was the perfect blend of emotion and hype." After the show, Winfrey held a press conference in which she claimed that the chemical companies needed to develop a "moral consciousness."[36]

In the summer of 1987, Miller and Leonard also worked together to plan a march to commemorate the third anniversary of the start of the lockout. On June 20, 1987, more than one thousand supporters of the union, including the majority of the locked-out workers, trekked through

Gonzales on a sweltering summer's day. They carried banners that read "No More Bhopals," "Danger BASF Toxic," and "3 Years of Foreign Oppression." The march again drew the support of environmentalists, who voiced their concerns about the safety of the plant. John O'Connor, director of the National Campaign against Toxics, told the crowd that both BASF and the Bhopal plant had a history of "lawlessness . . . and a lot of leaks." Marchers were also addressed by Amos Favorite and by OCAW vice-president Robert Wages. Attended by state senators, city councilmen, and several church leaders, the march highlighted the union's success in building allies in the community. Afterwards, participants enjoyed a jambalaya lunch cooked by locked-out workers.[37]

Events such as the lock-in and the anniversary march clearly illustrated the increasing amount of church support that Local 4-620 was now receiving. At the lock-in, for example, Reverend George Yunger of the American Baptists spoke publicly in support of the union's position. Dr. George E. Ogle, the program director of the United Methodist Church's Social and Economic Justice Department, also expressed his concerns about the company's actions on a number of occasions. Ogle insisted that BASF should make greater efforts to settle the dispute, and he refused to accept the company's arguments that it needed to change the seniority

18. The third-year anniversary march, June 1987. (Courtesy PACE International Union)

19. Amos Favorite addressing marchers at the third-year anniversary march.
(Courtesy PACE International Union)

system. "Those who have invested more of their lives in a firm should have some benefit over the newcomer," he told Les Story. "To do away with that system entirely seems patently unfair."[38]

It was the Roman Catholic Church, however, that was the most vocal in supporting the locked-out workers. Ernie Rousselle was a well-respected and widely known labor leader in Louisiana, as well as a keen Catholic, and he played an important role in securing the support of leading Catholics in the state. In addition, Leonard and Miller worked hard to inform church leaders of their struggle.[39] Stanley Ott, the Catholic bishop of Baton Rouge, was particularly supportive. On Labor Day in 1987, for example, he led a special mass at the Catholic church in Gonzales. In his homily, Ott prayed for the locked-out workers and their families and expressed his hope for a swift end to the lengthy dispute. While insisting that he was not "in favor of one side or the other," Ott noted that BASF had a duty to seek an end to the dispute, "especially since it was management action which first initiated the lockout of workers from jobs in which they had acquired certain rights." Citing Catholic doctrine that stretched back to Pope Leo XIII's 1891 encyclical *Rerum Novarum*, which had declared the right of workers to form unions, Ott reminded the congregation that these rights needed to be respected. He added that the needs of the replacement workers were "secondary to those locked out."[40]

Father George Lundy, director of Loyola University's Institute of Human Relations in New Orleans, spoke even more strongly than Ott in his support for the locked-out workers. In a sympathetic article on the lockout for Loyola's *Blueprint for Social Justice*, he described BASF as a "high profit, viciously anti-union corporation" that had a "terrible environmental record." Lundy praised the Geismar workers as "loyal seekers of the American dream for themselves and their families." Lundy also placed his support of the workers in the context of Catholic teaching, citing both *Rerum Novarum* and *Quadragesimo Anno*, which was issued by Pope Pius XI in 1931. *Quadragesimo Anno* went a step further than *Rerum Novarum*, adding that the church had a duty to encourage "Christian workingmen to form unions according to their several trades, and of teaching them how to do it."[41]

The fact that the BASF workers had been locked out clearly helped them gain church support. Asserting that Local 4-620's members were being unfairly deprived of their livelihood, several religious leaders asserted that offensive lockouts were morally wrong. In his correspondence, Lundy himself claimed that "Locking out 390 workers who want to continue working under their old contract is outrageous by any standard—a trampling of the most fundamental dignity and rights of whole families whose lives are completely disrupted." Sister Fara Impastato, an associate professor of religious studies at Loyola, described the locked-out workers as "true heroes" who were "fighting to claim for the workers of our country that dignity and freedom the mindless and heartless managers of things would like to destroy." Impastato, an outspoken critic of the German company, also felt that BASF was morally wrong to lock out the workers. Appearing on the union's *Locked Out!* video, she claimed that BASF had acted with "short sighted destructiveness" and was "very evil."[42]

The support of the Roman Catholic Church undoubtedly raised the morale of the locked-out workers, a majority of whom were Catholics. In a letter to Ott, Esnard Gremillion explained that the bishop's support had helped the union's members to feel that they were fighting for a just cause: "I am very glad that you are willing to take part in our struggle with BASF, because it means so much to me and to the members of OCAW Local 4-620. . . . We know final victory will be ours and we can return to work with the dignity God assures us." Many rank-and-file workers claimed that the support of the church was very important. "The Bishop was involved, and that was a comfort to a lot of people," remembered Carey Hawkins. ". . . It is comforting when you know that the mass, that the priest is making the comment that we will pray for the locked-out workers and their families. You know you're not forgotten."[43]

Church support was clearly important in helping many workers to deal with the financial and emotional effects of the lockout. By the third year of the dispute, however, two workers nevertheless found this pressure too much to cope with. The lockout had a devastating financial impact on Robert Washington, who had worked at the plant since it had opened in 1958. Working three janitorial jobs and heavily in debt, Washington committed suicide in June 1987 because he felt unable to cope. A fifty-one-year-old African-American, Washington was a well-known worker who had played a leading role in efforts to improve job opportunities for black workers at the plant. His death touched many other locked-out workers. Viewing him as a "fallen soldier," Local 4-620 members attended his funeral in large numbers.[44]

Some locked-out workers found that they had to travel many miles in order to find work. For those who had been used to a steady income from a secure job, this new transient life could be stressful and depressing. Jimmie McPhie, a maintenance mechanic before the lockout, took his camper van to Phoenix, Arizona, in search for work. In October 1987, the fifty-one-year-old worker was found dead in the vehicle from carbon monoxide poisoning. The coroner ruled the death "accidental," but locked-out workers universally viewed it as suicide, pointing out that the search for work had left McPhie severely depressed. Due to the length of the dispute, several other workers also passed away while it was going on, including Neil Cudd, a forty-five-year-old maintenance electrician who succumbed to cancer. The deaths of these workers influenced union members to continue fighting until they had a signed contract that allowed all of their members to return the plant. "They won't be forgotten," pledged Gremillion.[45]

As Miller and Leonard forged greater links with environmental groups, most locked-out workers experienced sustained personal contact with environmentalists for the first time in their lives. Both OCAW strategists tried to involve the rank and file in the campaign as much as possible, and workers and environmentalists had joined together to participate in both the lock-in and the anniversary march. In the spring of 1987, the two men had also organized an "Environmental March" to the Governor's Mansion in Baton Rouge. Locked-out workers, environmentalists, and local residents marched through the streets of Baton Rouge carrying signs that read "No More Permits," and "BASF, Bhopal on the Bayou?" Several locked-out workers also went on trips to Germany with environmental activists such as Amos Favorite, Darryl Malek-Wiley, and Willie Fontenot, while others came into contact with them by attending permit hearings in Louisiana. Workers who met environmentalists admitted that the experience caused them to change their views.[46]

Roy Fink, who had worked at the plant since 1965, summed up the way that personal contact with environmentalists began to transform his beliefs. "At first we were kind of leery of it," he acknowledged, "because all we'd ever heard, I guess the key people we'd ever heard of was Greenpeace, and really our concept was, 'They want to do away with all the chemicals in the world and close all these plants down.' That was really what our outlook was. When you talked about environmentalists, we kind of started cringing." Most workers had never met environmentalists before, however, and Fink recalled that their views changed once they did: "When we started out and we held a few meetings and stuff with a few of the environmental people and started finding out, 'Hey, these people are not really all that bad. They're after the same things we're after. They recognize the fact that you can make chemicals, you can deal with environmental hazards, but it costs more money. But you can deal with it and you can control the waste. You don't have to just go and dump it in the river or just dump it in the ground.' Like I say, the more we got involved with them, the more we found that there was environmental groups that wasn't what we thought they was. I mean we thought they were monsters, and even so far as Greenpeace, we've went in with some demonstrations with Greenpeace. They're a very intelligent bunch of people."[47]

Other workers admitted that they became more sympathetic to working with environmentalists when they saw that the environmental campaign was exerting some leverage against the company. "At first I couldn't see it really," admitted Putsy Braud. "I mean I didn't see how that was going to do us any good at all, but it did, no doubt it did. I had given up. I didn't think there was anything that could put the pressure on that company that it needed. . . . but this environmental put some pressure on them. It made them bend down and talk to us." Braud remembered that his awareness of environmental problems increased as a result. "It made everybody more aware of pollution," he acknowledged.[48]

Throughout the course of 1987, the OCAW's proposal that the lockout should be resolved by a citizens' review panel gathered momentum.

20. Roy Fink, pictured during the lockout. (Courtesy PACE International Union)

Originally proposed by Governor Edwards, the panel was conceived as a five-member body with management and labor appointing two representatives each, while a fifth representative would be selected by the governor. By the summer of 1987, the appointment of the panel was favored by a wide variety of prominent figures in Louisiana, including State Senators Joseph Sevario and Mike Cross, the Methodist bishop Ben Oliphant, the Gonzales Board of Aldermen, and the Sorrento Town Council. BASF, however, consistently refused to back the idea. It argued that the panel would be counterproductive, that it would raise false hopes, and that it would take a long time to educate members on the "complex issues" involved in the dispute. The company also pointed out that the panel's findings would not have the force of law.[49]

Although Edwards gave his support to the setting up of a fact-finding commission, BASF executives worked hard to ensure that he did not give open support to the union. Managers were aware that Edwards had traditionally received union support and had also been favorable to industrial development. They therefore stressed to the governor that they were not antiunion and argued that the OCAW's campaign was damaging the state's efforts to attract new industry. In one lengthy letter, Les Story complained to Edwards that the company was the victim of a "malicious campaign" conducted by out-of-state union activists. He suggested that this campaign could hinder the company's plans to expand the site: "Over the past two years our company has been investing $50 million in new facilities at the Geismar works. These new facilities are creating hundreds of construction jobs and about 50 high-paying full-time positions. We are hopeful more new facilities and jobs are on the way. However, the union's corporate campaign has put our State in a very poor light and I believe that severe damage has already been done to our State's reputation for investments."[50]

Although he received correspondence from both sides, Edwards continued to remain largely neutral in the dispute. While he supported the fact-finding proposal, the governor never intervened more decisively in the conflict, as BASF executives feared he might. As such, their lobbying efforts appear to have had some success.[51]

By the summer of 1987, however, the union's coalition-building strategy was clearly of concern to BASF. One executive noted in July 1987 that the union had achieved "some success" in its efforts to attract broader support and warned that BASF was in danger of being forced "into a difficult defensive position." Managers worried, in particular, that if the union continued to receive support from church and civic leaders it would push influential state officials to side with the locked-out workers. The company was urged by its public relations consultants to go on the offensive against the OCAW: "Sitting back and waiting may not be an appropriate strategy at this time."[52] Consultant Rafael Bermudez stressed that the company had to respond to an escalating corporate campaign: "My fear is that with so much activity going on, the locked-out workers could become a cause celebre and more parties will join the Union; thus, increasing the pressure on BASF." Apparently implementing this advice, company officials arranged a meeting with the Reverend James C. Carter, president of Loyola University, to try and stem further expressions of support from Loyola faculty for the union. Carter, however, made it clear that faculty members were given considerable leeway to express their own opinions but were instructed that when they were discussing public issues they were not speaking for the university.[53]

In response to the union's escalating campaign, BASF was also keen to put its own views forward. On a local level, the company's chief spokesperson was Les Story, and he continued to vigorously defend BASF's conduct. Story insisted that the company was being unfairly harassed by disreputable union leaders. He stressed the contribution that the company made to the local economy, contrasting this with the "professional out-of-state union organizers" who had come to the area simply to pressure BASF into settling the dispute on the union's terms. "Orchestrating this smear campaign is one Richard Miller, who recently traveled to Louisiana and has apparently taken up temporary residence in a motel on Airline Highway," noted Story. "Also assisting in this effort is another non-Louisiana resident, Richard Leonard. Both of these men are paid salaries and expenses by OCAW while the rank-and-file members of OCAW Local 4-620 remain without jobs."[54]

Story repeatedly asserted that the campaign was actually prolonging the dispute and that the union should devote its energies to finding a settlement. Picturing the union as the obstacle to an agreement, he asserted that OCAW workers were refusing the chance to operate their well-paid jobs. "The issues of this labor dispute are numerous and complicated," he told the *Shreveport Times*, "but the bottom line is that BASF has offered union members wages of $36,000 a year—an offer which they have rejected." As such, Story aimed to switch the focus of the dispute onto economic issues, avoiding the issue of job security that locked-out workers felt was at the heart of the conflict.[55]

BASF also argued that environmental groups were being "used" by the union to exert leverage against the company. The company attacked the APRATP, in particular, insisting that it was "not organized for legitimate environmental purposes but to harass and discredit BASF Corporation as part of an ongoing campaign."[56] The union had certainly helped the APRATP to organize, and Richard Miller, in retrospective interviews, did not deny that he was seeking to exert economic pressure on BASF by disrupting production at Geismar. The alliance with the union was beneficial to the parish residents, however, and both groups shared a desire to make the company more accountable. Amos Favorite himself rejected the accusation that the Geismar residents were being used. Hitting back at Story, he pointed out that the union was a genuine help to the residents of

Geismar. "We welcomed the help from the BASF workers because they have helped us out," he wrote. "We are painted as being used, when the reality is that we are actually using them to make our situation better here in Geismar." Favorite formed genuine and lasting friendships with many of the locked-out workers, and he found that their knowledge of the plant's environmental history was a great help to the APRATP. The union was indeed able to provide the residents with technical environmental specialists that the group could not have easily afforded on its own.[57]

Other environmental-organization leaders asserted that they benefited from working with the union. "We are finding," commented Darryl Malek-Wiley, "that the people who work in the plant, or are employed in these plants, have information that will help us." He rejected the idea that the relationship between workers and environmentalists was inherently conflictive: "They are helping us, we are helping them and we are all getting together. We all want the same thing—we want safe chemical plants, we want safe communities and we want very healthy families."[58]

Publicly, OCAW leaders hotly rejected charges that they were working with environmental groups simply as a means of gaining leverage against BASF.[59] Privately, however, it is clear that union strategists saw their bond with environmentalists primarily as an effective way of putting pressure on the company. In February 1987, Leonard wrote OCAW general counsel John McKendree to highlight the value of working with environmental groups: "I am becoming increasingly impressed with the way in which our good relationship with environmentally concerned citizens and activists can be used as a potent weapon/deterrent against unreasonable action by employers." Leonard noted that many environmental groups were interested in launching corporate campaigns but were unable to do so because they lacked the inside knowledge that union members could provide. Working together, the two groups could conduct a "truly lethal campaign" that he was confident would eventually bring the company back to the table. The fact that the union looked to the environmental alliance as a way of exerting leverage against BASF was, of course, no real surprise; the union's primary responsibility was to its locked-out members, and it needed above all else to find a way to return them to their jobs.[60]

By the fall of 1987, the campaign had clearly influenced BASF to seek a settlement. By this time, other parts of the chemical industry were expressing concern about the OCAW's activities. Four days before BASF lifted the lockout, for example, Esnard Gremillion met with Fred Loy of the LCA and Dow Chemical Company chairman Bob Gallant. Loy noted that the chemical industry was very concerned about the union's campaign. "OCAW should limit the fight with BASF and not include all chemical plants," he commented. "After all there are some 56 plants being affected by this bad publicity and (it) could cost jobs and discourage future growth of present facilities and keep new industry out of Louisiana." The union's activities, Loy added, were "giving the State a bad image." The LCA leader twice asked Gremillion whether the union would end its campaign if the dispute was settled. Gremillion refused to give this commitment, although he made it clear that the main goal of the union was to get their members back to work.[61]

BASF's internal documents indicate that by the summer of 1987 managers were keen to try and bring an end to the lengthy dispute, which was referred to in company sources as "the problem." In August 1987, for example, Parsippany manager Dave Buchner wrote that management wanted "to end the stalemate. BASF does not want to have to defend having the longest continuing lock-out in the United States."[62] Company officials had anticipated that the lockout would be shortlived, and by the summer of 1987 the length of the dispute was of real concern to them. As the longest-running lockout in the country, the company worried that it would attract media attention and provide "an instant 'angle' for any story."[63] Managers fretted, in particular, that the continuing lockout was damaging their ability to expand the Geismar site, which was described as "the most desirable for many new projects as well as for the existing operations." This confirmed that BASF regarded the Geismar site as extremely valuable and was thus keen to end the dispute so that the company could continue to expand it.[64]

The decision to lift the lockout was made jointly by BASF's managers in Parsippany and Geismar site management. As such, it reflected the company's traditional policy of allowing the Americans to decide labor relations strategy for American operations. Written records fail to provide any evidence of pressure from Germany, although it is clear that there were differences in approach between BASF's European and North American executives, and that the lockout exacerbated these differences.

While Hans Albers publicly claimed that BASF's German headquarters was continuing to allow U.S. managers to decide labor relations policies, behind the scenes German managers were not as distant from the dispute as he suggested. BASF's American management stressed that the lockout made German management very uncomfortable. The Americans

asserted that their German counterparts were more used to the coopera-
tive basis of German labor relations and had always disapproved of the
lockout. Richard Donaldson, for example, stressed that German management was "very very uncomfortable with this whole thing. This was just
something that didn't happen. If nothing else it was, I don't want to call it
bad manners, but they didn't do things like that in Germany. They didn't
treat their employees like that, you know, lock them out. . . . They grew up
with the unions that were in Ludwigshafen, so they know they're not bad
people, so what is this thing all about? They were very uncomfortable
with that whole thing. They didn't understand why in the world this was
going on."[65]

Many historians, together with other observers, have commented that
modern German society has a longing for harmony and consensus and is
fearful of conflict. Volker R. Berghahn and Detlev Karsten point out in
their study of West German industrial relations that industrial conflict
was "not accepted as something normal," adding that public opinion generally opposed the side that had violated the peace.[66] After the upheavals
of the 1930s and World War II, both management and labor in West Germany sought to build a secure and peaceful relationship. To West Germans, the lockout weapon was, as Berghahn and Karsten put it, "highly
controversial."[67] Employers' right to lock out workers was restricted in
West Germany, with offensive lockouts being illegal. In addition, all lockouts were illegal according to the constitution of the state of Hesse. In the
Federal Republic as a whole, defensive lockouts were allowed in response
to selective strikes, but such lockouts were intended to end the strike and
bring the union back to the bargaining table, and they were not allowed to
be "excessive" in length. Such a restrictive lockout was a far cry from the
extended and bitter dispute taking place in Louisiana.[68]

Bill Jenkins related that the lockout brought to light cultural differences between the company's management in the United States and Germany. From an early stage, he recalled that he had to do a lot of explaining
to the Germans, who were intrigued by the confrontational nature of the
Geismar lockout: "We were getting a lot of inquiries: Why are you doing
this? Why are you doing that? There was a lot of time required to explain
to Germany, and I think I shared with you I got a crisis call one time from
Stenzel. He said, 'Bill, you have to go to Germany. They want you to
conduct a press conference with the German press. The German press
has a lot of questions about the work stoppage and the lockout and why
we're doing this and is it legal and all this sort of thing.' It was kind of

interesting. . . . The German management had become concerned or they wouldn't have demanded that I go over and hold a press conference to explain what was going on, because they couldn't explain it." Like Donaldson, Jenkins felt that Germans found it hard to understood the confrontational nature of labor relations in the United States. "When they saw this problem," he reflected, "they just couldn't comprehend, and it was uncomfortable for them. . . . I don't think they understood American labor relations. You remember we talked earlier that they were heavily unionized in Germany, but the unions there, it's more like a family, it's not confrontational, and if there's an issue or a problem between management and the unions, it gets worked out. At Ludwigshafen, that's the way it had always been."[69]

The German managers were indeed much less accustomed to conflict with unions than their American counterparts. In the late 1980s, one high-ranking BASF German executive admitted that the lockout surprised managers who were used to labor peace: "We haven't had a strike since 1924, except a work stoppage in 1947 to protest our president being tried for war crimes."[70] In one interview conducted by West German public broadcasting, these differences were also apparent, as correspondent Hermann Vinke spent a great deal of time asking Les Story to explain why he had locked the workers out. The move, Vinke asserted, was "beyond the normal level of, well, differences between management and workers."[71]

All of the top American managers active at the time of the dispute presented this same picture of German managers being very concerned and puzzled by the dispute. Publicly, BASF's European headquarters refused to confirm this, insisting that the company had a united strategy. "There was no difference of approach," insisted Nicola Palmieri-Egger, an Italian-born BASF executive.[72] As early as October 1986, however, Palmieri-Egger himself sent a memo to the company's U.S. headquarters that expressed some concerns about the actions of American management. "BASF has a worldwide image to protect," he noted. "Taking good care of its employees is part of it." In response, U.S. executive Dave Buchner justified the company's position and noted that lockouts of this type were not "unusual and abnormal" in America.[73]

The concern of German managers about the dispute is clear, and it is possible that they may have exerted informal pressure to settle the dispute. Both Henry Kramer and Bill Jenkins certainly felt that BASF's German managers were also keen to lift the lockout. As corporate manager of labor relations, Kramer stressed that he had to spend "considerable time" trying to pacify European executives "who did not understand how labor relations work in the U.S. and had the German pejorative idea of a lockout."[74] Nevertheless, with the dispute disrupting their plans to expand Geismar, many American managers also wanted to try and bring an end to it by this time. Kramer, who was not in favor of any compromise with the union, had become increasingly isolated within BASF's labor relations team. Disagreeing with the decision to lift the lockout, he left the company at the end of 1987 and was replaced by John Kirkman, another Parsippany-based labor relations executive.[75]

As negotiations had reached an impasse, managers knew that they could lift the lockout by implementing their contract offer, as was allowed under U.S. labor law. Perhaps keen to make a fresh start with the union, BASF also promoted Les Story to a position in Parsippany and brought in a new site manager from their Freeport, Texas, plant. It is possible that Parsippany saw Story as an obstacle to a settlement, although Story himself asserted that he was never told that this was the reason for his transfer. In addition, Story continued to be involved in the Geismar negotiations even after he had moved to New Jersey. Some salaried workers in Geismar did feel, however, that Story was removed so as to clear the way for a settlement.[76]

On September 22, 1987, BASF sent the OCAW a complete offer for a new collective-bargaining agreement. The agreement included the "total subcontracting" of maintenance work at Geismar, with the replaced maintenance workers having no recall or seniority rights. The union was given until October 27, 1987, to respond; once this date elapsed, the lockout was lifted for the operators. The company proposed that the displaced maintenance workers be given severance pay. The union, however, refused to discuss this, asserting that the workers were not being replaced. BASF stressed that the maintenance workers were being offered severance payments of up to $25,000 each, although the union asserted that the average payment was around $10,000. Once the company had implemented its offer, the OCAW filed unfair labor practice charges against BASF, asserting that the company had bargained in bad faith and that its decision to subcontract maintenance was not based on a desire for efficiency but was merely a ploy to destroy the union.[77]

Under the implemented offer, the union's rights were strictly curtailed. BASF suspended the checkoff of union dues, ensuring that workers had to pay them voluntarily. The grievance procedure was also simplified, with the outside arbitration of grievances being abolished. Workers therefore had no redress if BASF denied their grievances. Local 4-620 operators would work under these provisions until December 1989, when the two sides finally agreed a new contract. Despite these limitations, the company's offer was seen by the union as a sign that BASF was growing weary of the dispute. Miller and Leonard remained upbeat, confident that they could continue to recruit allies in the community and exert enough pressure on BASF to secure the return of all of Local 4-620's members to the plant.[78]

It Ain't Over

In lifting the lockout, BASF executives had clearly hoped that their dispute with the OCAW would become less intense. Over the course of 1988, however, the union continued to escalate its corporate campaign and to work increasingly with environmental groups. As Richard Leonard noted, the union viewed its battle with BASF as a "war" that it must continue to fight until the company signed a new contract that allowed all the workers who had been locked out the opportunity to return to the plant.[1]

The union's reaction to the implemented offer highlighted that the company's hopes of an end to the dispute would not be realized. As the maintenance workers were not being given the opportunity to return, Local 4-620's members intended to carry on fighting the company. Esnard Gremillion summed up the union's position: "It's clear that the lockout is not over. While the company may be calling back some operators, the maintenance workers will remain locked out." At a union meeting in October 1987, more than three hundred local union members publicly informed BASF of their determination to fight on by holding up signs that read "It Ain't Over." The union's position was supported by several of the union's newfound allies. Bishop Ott, for example, noted that "the labor dispute is far from over . . . a substantial number of workers effectively remain 'locked out.'" Louisiana attorney general William Guste also criticized the company's offer, asserting that workers should be recalled en masse.[2]

The return of the OCAW operators ensured that they would have to work alongside the contract operators in the plant. Although these workers had only been hired on a temporary basis, the continued expansion of the plant allowed the company to retain them, with many eventually securing permanent jobs. Managers were also keen to keep these workers because they were aware that laying them off might give some credence to the union's contention that they were not as highly qualified as the

21. Local 4-620 members express their determination to fight on following the return of the operators, October 1987. (Courtesy PACE International Union)

OCAW operators.[3] Not surprisingly, the relationship between the two groups was often poor. "There were words exchanged in a lot of places," admitted Bobby Schneider, who acted as the local's "inside" chairperson after returning to the plant.[4]

Carey Hawkins acknowledged that returning to work was "very difficult" because he felt a lot of resentment toward the contract operators. During the lockout, Hawkins had struggled to find a steady job and had ended up in court because of his failure to make child-support payments, an incident that he described as his personal "low point" of the dispute. Once inside the plant, he faced a difficult personal battle to try and contain his anger toward the contractors. "I try, I do pray to be more tolerant almost daily and to not hold grudges," he reflected more than a decade after the dispute had ended. "So the best thing for me was to just stay away from them as much as possible, and that was the way I coped with it, because, yeah, I had a lot of hard feelings. It was like, 'You didn't go to court because you couldn't pay your child support payment, I went. You've been having this job.'"[5]

The union and the company disagreed about how the OCAW members were to be recalled. Local 4-620 argued that the company should follow seniority in bringing back the operators, with the most experienced given the first opportunities. BASF managers flatly rejected this, asserting that they had to staff the site according to production needs and could not follow seniority. "The union wanted us to bring people in in seniority order, and we said, 'No, we weren't going to do that,'" recalled Richard Donaldson. For union members, the company's refusal to follow seniority only confirmed their long-standing belief that executives wanted complete freedom to staff the plant without interference from the union.[6]

Conflict between the company and the union continued after the union operators were recalled. Throughout 1988, Local 4-620 members repeatedly claimed that the company was discriminating against them in the assignment of overtime and in promotions. In November 1988, for example, nine workers in the glycol department filed a joint grievance complaining that contract workers were being promoted ahead of the recalled union operators. These complaints continued right up until the company and union finally agreed a new contract. In the course of 1989, for example, a group of OCAW members in the polyol department complained that contractors were receiving more overtime than they were, while others alleged that the company was not following established shift rules. While some grievances were genuine, many also reflected union members' desire to continue fighting BASF. With no arbitration available to Local 4-620 members, however, BASF was able to flatly deny all of these grievances.[7]

At the negotiating table, the company and union were unable to make progress on the few occasions when they did meet. Throughout 1988, in fact, the relationship between the two sides was so poor that they were often unable to even agree on a meeting time or place. When they did get together, the subcontracting of maintenance continued to be the main stumbling-block. BASF repeatedly refused union demands to document the superior efficiency of the contract maintenance workers, insisting that the company's decision to subcontract "was not based upon documentation, but rather upon its experience with subcontract maintenance." The union, for its part, stubbornly refused to consider any offers by the company unless they included job security for all of Local 4-620's members, which they did not.[8]

Many union members felt that BASF's tactics were designed to divide the union. "They tried to divide us by bringing the operators in and keeping the maintenance people out," asserted Bobby Schneider. "They figured that once we got in there and all, they would be able to get it that way. Well that didn't work." Feeling that BASF was trying to split the local union, operators were particularly determined to resist BASF's offers until all the maintenance workers had been given the opportunity to return to work. Now that the automatic checkoff of union dues had been removed by BASF, most workers were also determined to prove their union allegiance by paying their dues voluntarily, although a small number did fall behind in their payments.[9]

While operators were returning to the plant, the international union continued its efforts to try and create political pressure against BASF. In March 1988, the lockout was investigated by the U.S. House Subcommittee on Labor-Management Relations. The committee heard union allegations that companies were increasingly using offensive lockouts as a way of destroying them. Maintenance worker Bodie Southall told the committee that BASF's main goal was to eliminate Local 4-620. "The issue here is whether or not a union and its members will continue to exist at Geismar," he asserted. After Southall, the committee heard testimony from other several other workers, including union members locked out of International Paper Company's Mobile, Alabama, plant. Between 1988 and 1995, as the AFL-CIO pushed hard for a strengthened National Labor Relations Act, there were at least nine separate congressional hearings that debated whether labor law reform was needed. These efforts to change the law were met by determined resistance from the business community. Claiming that labor law reform would give too much power to unions, employers and their supporters mobilized to ensure that AFL-CIO backed proposals were defeated by Republican-led filibusters in the Senate in the summers of 1992 and 1994.[10]

As had been the case earlier in the dispute, it was the union's environmental campaign that was to prove most effective at exerting pressure on BASF. Although few people realized it at the time, the election of Charles "Buddy" Roemer as governor of Louisiana was to help the Geismar workers. In the mid-1980s, a drop in oil and gas revenues had forced Edwin Edwards to push through an unpopular $730 million tax increase in an effort to stave off fiscal chaos. When this measure failed to be sufficient, Edwards was pushed to slash government services, creating further voter discontent. A congressman since 1980, Roemer benefited from this mood of discontent and defeated Edwards in the first primary in 1987, pushing his opponent to concede defeat and back out of a second primary.[11]

Coming into office in March 1988, Roemer promised a revolution in Louisiana politics that would free the state from corruption.[12] He was a conservative Democrat, however, and OCAW leaders initially worried about his election. Organized labor had traditionally been a strong supporter of Edwards. Although Edwards had remained largely neutral in the

BASF dispute, he had supported the setting up of a fact-finding committee, a proposal favored by the union but vigorously opposed by BASF. Roemer, in contrast, refused to endorse the proposal, mainly because the state AFL-CIO had solidly backed his opponent in the election.[13]

Roemer's election would ultimately help the union, however. Arguing that his predecessor had allowed companies an unnecessary amount of leeway to pollute, Roemer introduced much tougher environmental standards. The new governor traced his environmental concern to his upbringing in rural northern Louisiana, away from the heavily industrialized chemical corridor: "I'm the first governor from the north, and I brought with me that childhood belief in the land and the water. . . . I was free from the BASFs, from the Exxons, from the big companies I was free, and I maintained with pride my freedom."[14] To head a revitalized DEQ, Roemer hired Dr. Paul Templet, an environmental studies professor at Louisiana State University. Backed by a major budget increase, the DEQ under Templet became much more rigorous in its enforcement of environmental standards. Realizing that corporations disliked negative publicity, Templet's DEQ published *Corporate Response*, a list of the biggest polluters in the state. Many companies placed on the list took prompt action to remove themselves from it. Together with other initiatives, including a reformed permit-application process, *Corporate Response* succeeded in cutting the state's toxic releases from 840 million pounds in 1987 to 482 million pounds in 1989.[15]

Executives, not surprisingly, disliked Templet and pushed hard for Edwards' reelection in 1991. LABI president Ed Steimel admitted that Templet was "hated by the chemical industry." At the same time, Steimel accepted that Templet's efforts had cleaned up the environment: "I think he was very sincere in what he was doing, and he is responsible for an awful lot of the improvement in the environment, there's no doubt about that. . . . I think Paul Templet did us all a big favor."[16]

At the time of Templet's appointment, the union had joined with local environmental groups to launch a major drive to try and stop BASF from injecting waste into the ground at the Geismar site. In the 1980s, deep-well injection was a popular method of hazardous waste disposal in Louisiana; in 1988, for example, over 72 percent of the state's hazardous waste was injected underground. The method was preferred mainly because it was much cheaper than alternatives such as biological degradation or incineration. With thirteen injection wells located within a six-mile radius of Geismar, local residents had genuine concerns about the amount of waste injected in the ground, especially as studies commissioned by the OCAW and the APRATP highlighted that the method could contaminate their drinking water. In 1986, union figures claimed that industries injected over 106 million gallons of liquid hazardous waste into the ground under Geismar.[17]

In 1984 the Louisiana legislature had passed Act 803, which required the DEQ secretary to identify wastes that were unsuitable for land disposal and deep-well injection and to set deadlines for prohibiting their disposal by these methods. The secretary was also required to find feasible alternatives to deep-well injection and to permit it only as a last resort. Despite this, however, the DEQ failed to meet the January 1, 1986, deadline to set up the new regulations and continued to allow waste to be injected. In total, twenty-one deep-well permits were issued by the agency between January 1986 and February 1988, including a permit for BASF to inject hydrochloric acid (HCl). "Rather than creating and following guidelines which protect public health, DEQ has been rubber stamping industry requests to inject hazardous wastes," charged Amos Favorite.[18]

Seeking to defend the injection of HCl, BASF repeatedly argued that it lacked viable alternatives to dispose of the waste. The company claimed that it already used two other disposal methods—selling or reusing the HCl—to "the maximum extent feasible." Other possibilities, such as transporting the waste off-site for subsurface disposal, were rejected by BASF from both an economic and environmental standpoint.[19]

These arguments failed to assuage the concerns of many residents, and the union clearly tapped into public concern in its fight against waste injection. At the start of 1988, more than four hundred people attended a hearing in Gonzales to debate the merits of BASF's permit application to dispose of 31,000 pounds of waste a day by underground injection. In addition to union speakers, local residents and environmental leaders harangued the company, trading catcalls and boos with industry supporters. Hitting back at industry claims that "the technology used today is state-of-the-art," Plaquemine environmentalist Les Ann Kirkland claimed that "waste injection is not a technology; it's a hole in the ground." Local resident Tammy Guillory refused to be cowed by industry claims that the chemical plants created jobs. "There won't be any jobs and no money without our health," she asserted. Locked-out worker Darryl Stevens supported the residents, claiming that injection wells were "potential time bombs planted underground" that threatened Geismar residents' drinking water.[20]

In early 1988, the OCAW and the APRATP stepped up their fight by filing suit in East Baton Rouge Parish District Court to shut down the state's deep-well-injection permitting program until the DEQ promulgated regulations to limit the injection of hazardous waste. They were assisted by the new Tulane Environmental Law Clinic, which represented a variety of state environmental groups, including the Sierra Club and LEAN. The litigation was largely successful; in March 1988, Judge William H. Brown ordered the DEQ to promulgate the hazardous waste regulations required by Act 803, giving them four months to complete the regulatory process. Facing considerable pressure from environmental groups, the DEQ also agreed to issue no more well certifications until the regulations were promulgated.[21]

The deep-well-injection fight illustrated the way that Richard Miller aimed to improve the environmental practices of the chemical industry as a whole and not just of BASF. Aiming to build community support, Miller sought new regulations for the entire state. As a union press release on the issue put it: "While the issues appear to target BASF, they are designed to challenge the very foundations of an under-regulated industry and an inadequately administered system of law."[22] The joint OCAW-APRATP litigation certainly slowed up BASF's permit application. The company was still seeking the permit when it finally signed a new contract with the union in December 1989, more than three years after making its initial application. Ongoing concern about whether waste from the well would migrate into residents' drinking supplies was the main cause of this delay. In the past, by contrast, such permits had generally been routinely granted.[23] Drawing on research he had commissioned from hydrologists, Miller continually peppered the DEQ with detailed objections to deep-well injection, even writing letters on public holidays.[24]

Most of Miller's activities, however, were still focused on BASF. By the summer of 1988, the OCAW strategist had extensive contacts within the local area, and he used these to act as a vocal whistle-blower about any potential environmental malpractice by the German chemical company. His letters to DEQ officials were detailed and showed a close knowledge of state environmental laws. In October 1988, for example, Miller asked the DEQ's enforcement division to investigate two kerosene spills on the BASF site. "Section 9.1.1 of the Notification regulations of the DEQ requires the company to report within 24 hours upon discovery of the spill," he noted. "This was not done by BASF and therefore constitutes a violation of DEQ regulations. Will your office be carrying out an enforcement action against BASF for this violation? Will a penalty be assessed?"[25] In June 1988, Miller called for separate public hearings regarding an injection well "blowout" and unpermitted treatment of hazardous waste, while in August he spoke out against three separate permits that the company had applied for.[26]

New DEQ secretary Paul Templet wanted to make the agency more responsive to citizens' needs, and he took all of Miller's correspondence seriously. Templet indeed looked back on Miller as a valuable source of information and praised his effectiveness. "He certainly got our attention, and he dealt from a basis of fact," recalled Templet. "He wasn't just spouting words at us or blowing smoke-screens, like a lot of the industry guys did. He had facts at his fingertips, so we listened, and we did verify it, whether or not his information was correct, and when it was, we used it." Overall, Templet felt that the union's environmental campaign was "extremely effective."[27] He called the union "the public expert" on pollution, adding that, "they had their own technical people. They prepared papers and gave them to us. It made us very careful about how we dealt with the company knowing there were technical experts on the union side."[28]

Buddy Roemer agreed with Templet's assessment. After losing the 1991 election, Roemer retired from politics and became a businessman in Baton Rouge. Looking back on the lockout, Roemer stressed that the union effectively capitalized on the more environmental stance that his administration had taken: "The union was wise and wily in their use of environmental issues. I don't know what their beliefs were, but their tactics were very much pro-environment, because they knew we were sympathetic. They knew the DEQ could be an ally at bringing the company to accountability."[29]

The change in personnel at the DEQ clearly provided a major boost to the union's environmental campaign. In December 1988, Geismar site manager William C. "Bill" Moran himself noted that the relationship between his company and the state agency was changing. "There are several new DEQ appointees who believe industry has not been environmentally responsive and who believe past DEQ administrations have been ineffective," he warned another manager. "The attitude of these people toward BASF and other environmentally responsive companies will gradually change." Moran worried that Templet's administration was likely to refuse permits for hazardous waste injection into deep wells, admitting that this was a boost to the union's campaign: "In this case, OCAW may have jumped on a potentially successful bandwagon."[30]

Miller's correspondence with the DEQ pushed Moran to write Templet in October 1988. In a detailed letter, he claimed that the OCAW strategist was using the state's environmental laws to aid him in a "campaign of harassment, intimidation and calumny against BASF." Moran asserted that "Mr. Miller and OCAW are trying to force BASF to capitulate to their demands through various tactics, the main one at present being harassment through Louisiana's environmental regulations." The company tried to steer the dispute back to traditional labor-management issues, insisting that the environmental campaign was distracting the union from the central issues of the dispute and thus preventing a settlement. Chronicling all of the union's letters to the DEQ, Moran revealed once again just how closely BASF was monitoring the union's campaign, and that the company was genuinely concerned by these efforts.[31]

The company also tried to respond to the union's campaign by inviting community leaders to tour the Geismar plant. These plant tours did succeed in cementing the support of several figures, including Gonzales mayor Johnny Berthelot. The company continued to refuse tours to environmental groups, however, insisting that they would not listen to BASF's case. The union, for its part, saw these tours as a propaganda exercise and insisted that BASF should allow independent environmental experts into the plant. Asserting that the company had made special efforts to make the plant look superficially clean, Richard Leonard dismissed the exercise as "the Geismar Shiny Pipe Tour."[32]

In the spring of 1988, BASF announced that it wanted to build a paint plant, hazardous waste incinerator, and landfill at a new site in the Midwest. In all, the company promised to create up to three thousand permanent jobs in a $150 million expansion that it planned to locate either in Terre Haute, Indiana, Evansville, Indiana, or Haverhill, Ohio. Miller recognized that this major expansion provided him with a new opportunity to try and disrupt BASF's plans to increase its presence in North America. Together with locked-out workers, he visited all three communities and informed residents of the union's case against the German chemical-maker. BASF, he claimed, was a company with "a long history of environmental accidents" that "cared little" for worker safety.[33]

These visits to the Midwest were widely reported in the local press and certainly generated a great deal of negative publicity for BASF, which was often forced to respond to the union's accusations.[34] While in the Midwest, Miller also helped to organize opposition to the company. In Indiana, he hired outspoken local environmentalist John Blair as a policy consultant. Blair emerged as a leader within Citizens for a Clean County (CCC) and Valley Watch, environmental groups that led opposition to BASF in Terre Haute and Evansville. Under Blair's leadership, both groups grew in strength and influence. When BASF first announced its interest in the Terre Haute area, for example, local government leaders, who were supportive of the company's plans, boasted that they faced no organized opposition. By January 1989, however, CCC's growth had surprised even its leaders. "It has had some kind of snowball effect," one commented. "I don't think anybody could have anticipated we would grow like this."[35]

Residents of all three communities were eager to learn more about the giant chemical-maker and showed a willingness to listen to and publicize the union's allegations. Locked-out workers, including Roger Arnold and Esnard Gremillion, traveled to several cities in the Midwest and presented their case at citizens' meetings. They told local people about the lockout and the environmental problems of "Cancer Alley." Some members of the public were swayed by the locked-out workers' experience. In Portsmouth, Ohio, for example, one woman commented: "Our area needs jobs, but not the kind that is being offered by BASF."[36] Other residents noted that they did not want their area to become another "Cancer Alley" and cited the company's environmental record as reasons for opposing BASF.[37] The local press frequently reported the union's activities. In March 1988, for example, the *Evansville Courier* ran a story that detailed OCAW's accusation that BASF had a poor environmental and labor record. Even if Evansville was selected, the *Courier* noted, "the corporation likely is to be hounded by some environmental and labor groups. They aren't convinced a BASF plant would be the panacea local leaders are hoping for."[38]

Throughout the controversy, BASF consistently concentrated on the economic benefits that its proposals offered. In the Terre Haute area, for example, the company mailed a flyer to residents that claimed that the new plants would bring "safe, secure, well-paying jobs" to an area that had recently suffered a string of major business closings.[39] These economic prospects lured city leaders, including the mayor of Terre Haute, into supporting BASF's proposal.[40]

The company also secured the support of many trade union members. "We're for whatever it takes to get us work," declared one member of the carpenter's union.[41] Many members of the building trades unions also supported BASF because they stood to benefit from the construction jobs that the expansion would generate. Richard Miller was most disappointed, however, by the fact that some OCAW members took the company's side. "There are plenty of EPA regulations to protect the area," claimed one OCAW member in the Haverhill area. "What we need are jobs to bring the area out of the backwoods."[42] The support of OCAW members for BASF threatened to undercut the union's campaign. "Of course," noted Miller, "it is difficult for OCAW to attack BASF's expansion plans in southern Ohio if our OCAW locals in the area are simultaneously attacking environmentalists and saying, 'these environmental kooks are trying to destroy our jobs.'"[43]

Miller worked hard to try and limit the damage caused by unions' support of BASF, securing a few positive results. Charles Whitlock, a vice-president of a United Steelworkers' local in the Haverhill area, claimed he would oppose BASF after hearing the union's case: "Even if it means 50,000 jobs, it's not worth it if the plant isn't going to be safe."[44] On balance, however, most union members in the area supported the company's plans.[45]

BASF later rejected Evansville as a possible location, citing a soil study that showed the area was unsuitable because it lacked the impermeable layer of clay that the company needed for its operations. Local environmentalists, however, claimed that public opposition had led to this decision.[46] BASF also lost interest in Haverhill after soil tests "found less clay than is desirable for an industrial landfill."[47] In Terre Haute, the CCC filed a petition for judicial review of the proposed expansion. In July 1989, Special Judge Frank Nardi ruled on the case and dealt a severe blow to BASF's plans when he declared that an economic development plan that had been approved by city officials was null and void. Nardi ruled that several officials who had passed the scheme had conflicts of interest, including a county redevelopment commissioner whose son owned land that was scheduled to be bought by BASF. As the *Sullivan Daily Times* noted, Nardi's decision sent BASF's plans "back to square one." By the time it eventually signed a contract with OCAW Local 4-620, BASF had been unable to select a site for its major expansion, partly because of the OCAW's opposition.[48]

As part on its ongoing effort to create international pressure against BASF, the OCAW also filed a complaint against the company with the Governing Body of the International Labor Organization (ILO). Based in Geneva, the ILO was created as a result of the Treaty of Versailles and was originally a part of the League of Nations. After World War II, it gained status as an official part of the United Nations and was recognized throughout the world as an authoritative body on worker rights and labor standards. The ILO Governing Body was a tripartite elected body composed of fifty-six titular members: twenty-eight governments, fourteen worker representatives, and fourteen employer representatives. Working with the AFL-CIO's Paris office, the OCAW filed a complaint with the ILO in February 1988. The complaint was filed against BASF and the U.S. government and cited a "pattern" of antiunion behavior by the German company at Geismar and other U.S. locations. It claimed that U.S. labor legislation had proven to be "inadequate" to protect the rights of the company's American workers. The ILO's Freedom of Association Committee conducted an investigation into the charges, producing a report that was subsequently adopted by the Governing Body.[49]

In May 1988, the OCAW received a major boost when the ILO report was released. It criticized BASF for its conduct in the lockout and called on the U.S. government to give union grievances a rapid hearing. The report also asserted that BASF's decision to subcontract the maintenance work had not been linked to economic necessity. Rather, the international body accepted the union's argument that BASF had singled out for subcontracting the department in which most of the local union's leaders worked. "Subcontracting accompanied by the dismissal of union leaders can constitute an act of anti-union discrimination just as dismissal, compulsory retirement, downgrading, transfers or blacklisting," concluded the report. Although the ILO called on the U.S. government to use its "national machinery" to hear the union's case, it lacked the power to compel a settlement to the dispute. In their response, BASF executives emphasized this and sought to downplay the impact of the ILO's ruling. Nevertheless, the ruling did help to discredit BASF, particularly in Europe, and it may have contributed to the long-standing unease that German executives felt about the company's ongoing conflict with the union.[50]

While many aspects of the union's campaign did secure results, relations between the OCAW and I.G. Chemie were still fraught with difficulties. In the spring of 1988, OCAW leaders were shocked and annoyed when a delegation of I.G. Chemie officials toured the Geismar plant without informing their American counterparts of their visit. While in Geismar, an I.G. Chemie leader also proclaimed that the lockout was over because the operators had been called back, a line of argument that the company was trying to promote. OCAW leaders were incensed. In a letter to I.G. Chemie secretary general Herman Rappe, OCAW president Joseph Misbrener called the remarks of the German delegate "inelegant and insensitive."[51] The OCAW broke rank for the first time after the incident by publicly criticizing I.G. Chemie. For the Americans, the episode highlighted that I.G. Chemie leaders were too willing to listen to BASF's version of events, an allegation that they made repeatedly. They complained that the German union was too bureaucratic and conservative and should not have been so supportive of the company over health and safety issues. The leaders of the German union were indeed reported to be "extremely uncomfortable" with the OCAW's environmental stance, causing complaints from the Americans.[52]

From the perspective of the OCAW's leaders, the state protection afforded the German union had robbed it of its grassroots militancy, making it unable to understand well the dispute in Geismar.[53] Lacking experience of conflict with the company, I.G. Chemie leaders seemed to find it impossible to accept that the union did not carry some blame for the lockout. Relations between the two organizations never fully recovered after this incident, and OCAW leaders even began to privately refer to the German union as "I.G.Crummie."[54]

Even at this time, it is also clear that I.G. Chemie leaders were still alienated by the earlier swastika demonstration. BASF managers in fact proved adept at exploiting the damage caused by the incident. As late as February 1988, for example, Leonard complained that the company was effectively dividing the German and American unions: "BASF has moved smartly by repeatedly referring to us as the Union who accuses BASF managers of being Nazis." The company, he added, "have used such statements to drive a wedge between OCAW and the German union."[55]

In the fall of 1988, Local 4-620 was a cosponsor of the "Great Louisiana Toxic March," a major protest organized by a coalition of environmental activists and citizens' groups. The idea for the march came out of a meeting between Pat Bryant, the director of the Gulf Coast Tenant Leadership Development Project (GCTLDP), Darryl Malek-Wiley, and John Liebman, a Greenpeace organizer. While discussing ways to focus attention on Louisiana's toxic pollution problems, the three men mapped out plans for a eighty-mile march along the chemical corridor from Baton Rouge to New Orleans. The trek was planned to coincide with the arrival of a Greenpeace boat that was traveling along the Mississippi to highlight the river's pollution problems.[56] Looking back, Malek-Wiley recalled the genesis of the march clearly: "We were meeting at Pat's house and drank a bottle of wine. John was telling us that Greenpeace was planning to have a

22. The OCAW joins with Greenpeace, the Gulf Coast Tenant Leadership Development Project, and the Sierra Club to cosponsor the "Great Louisiana Toxics March," November 1988. (Courtesy PACE International Union)

boat come down the Mississippi River in 1988, it would be real interesting if something happened along the way. So we just sort of kicked ideas around and decided we'll do a march. And we called it the Great Louisiana Toxic March."[57]

The protest drew inspiration from the civil rights movement, which had often used marches as a way of organizing local communities.[58] The organizers of the toxic march similarly hoped to use it as a way of increasing the environmental awareness of local citizens in the communities that they passed through along the way. As well as striving to protest against the discharge of cancer-causing chemicals into the air, water, and soil of Louisiana, the march aimed to "end the discriminatory practices of locating hazardous waste sites and chemical plants in economically depressed and minority neighborhoods."[59]

In addition to the Sierra Club and the GCTLDP, LEAN and OCAW Local 4-620 were the major sponsors of the march. Although the OCAW had worked with both LEAN and the Sierra Club before, the march was its first major collaboration with the GCTLDP. Organized in 1983, the GCTLDP fought for the needs of African-American tenants along the Gulf Coast, quickly becoming very concerned with the environmental problems faced by many predominantly black communities in Louisiana. Flyers sent out to publicize the march highlighted Local 4-620's sponsorship and put forward the union's argument that an organized workforce was safer because union workers had more rights to refuse to carry out unsafe practices. "Unions are part of the solution," the flyer asserted. ". . . The Louisiana Toxic March supports trade unions and opposes union busting."[60]

On November 11, 1988, around one hundred protesters gathered at Devil's Swamp, an abandoned toxic waste dump north of Baton Rouge, to begin their long trek. Carrying signs reading "Breath or Death," "Industry Clean Up Your Act—We Have to Live Here," and "Clean Air and Water—An Endangered Species," the protesters were told by Bryant that they were "working together for the long haul to assert our rights for clean air, clean land and clean water."[61] The next day, at an Environmental Fair held on the capitol steps, Martin Luther King III, the son of the slain civil rights leader, spoke of the convergence between the ideals of the civil rights movement and the environmental movement and called on more African-Americans to become involved in fighting the problems caused by toxic waste.[62]

On day three of the protest, marchers entered Reveilletown, a largely abandoned black community that stood in the shadow of a Georgia Gulf chemical plant. Former residents claimed that residing next to the plant had been a "living nightmare" in which they were forced to combat obnoxious odors and PVC dust. "They say the quality of life increases when industry moves in, but the quality of life decreased in our neighborhood," one commented. By the time of the march, most residents had used compensation payments from Georgia Gulf to move away, and the original community had become a ghost town of crumbling shotgun houses.[63] The marchers then proceeded to St. Gabriel and Geismar, where they protested in front of the BASF plant. Linking the issues of environmental safety and union rights, marchers called for BASF to end the lockout, echoing the union's argument that a unionized workforce would improve the plant's environmental safety record. Aware of the publicity surrounding the march, BASF managers supplied drinks and portable toilets for the marchers, but they steadfastly refused to discuss environmental issues.[64]

Just prior to entering New Orleans, the influence of the civil rights movement upon the protest was apparent as marchers prayed in front of Jefferson Parish sheriff Harry Lee, who demanded that they pay for a police escort through his parish. Despite exhortations from the marchers such as "Change him, Lord! Look in his eyes, Lord," Lee refused to moderate his position.[65] Along the route, march organizers welcomed citizens to join with them in walking short sections through their communities. On some sections, including the final leg into New Orleans, as many as two hundred people took part, although a smaller core group walked the whole way.[66]

The march was planned to coincide with Greenpeace protests in the state. For more than three months prior to the protest, Greenpeace activists aboard the *Beluga*, a converted fire-fighting vessel, had been traveling down the Mississippi River to try and publicize the river's pollution problems. While the march was proceeding through the state, members of the boat's crew succeeded in carrying out a series of eye-catching protests, including a two-day action in which four protesters chained themselves to a bridge over the Mississippi River, resulting in major traffic disruption on Interstate 10. The protesters also strung a large banner from the bridge that read "Cancer Alley, La."[67]

Locked-out workers and WSG members took part in some of these activities. Roger Arnold, for example, traveled out into the Mississippi River with Greenpeace activists and helped to hold up a large sign that read "Sportsman's Paradise or Toxics Wasteland?: The Choice Is Ours." Locked-out workers and WSG members, along with reporters, watched from the riverbank as the action took place. The slogan was a reference to the state license plate, which proclaimed Louisiana to be a "Sportsman's Paradise." The union also held a press conference with the environmental group in which the same banners were displayed.

While they were in Louisiana, Greenpeace activists also tried to block a discharge pipe from the Georgia Gulf plant in Plaquemine, although the company claimed that these efforts were unsuccessful. While some members of the environmental community in Louisiana tried to distance themselves from these tactics, they also recognized Greenpeace's ability to raise the profile of environmental issues in the state.[68]

Public reaction to the march, as recorded in editorials and letters in the press, was mixed. Some residents, especially Africans-Americans who lived in the shadow of chemical plants, welcomed the attention that the protests were focusing on environmental problems. Several African-American political leaders also came out in open support of the march, including State Representatives Kip Holden and Avery Alexander.[69] The march certainly succeeded in raising public awareness of the "Cancer Alley" controversy.[70] Some observers, particularly column writers, opposed the protest and claimed that the "Cancer Alley" slogan would damage Louisiana's economy and image. "Can you imagine what the term 'cancer alley' does to real estate sales, which in turn affect the area's merchants, which in turn affect sales taxes and property taxes, which in turn affect the public bodies' operations and so on and so on," asked one concerned columnist.[71]

Looking back, Malek-Wiley was understandably keen to stress the significance of the protest, especially the fact that it was organized by a biracial coalition. A critic of what he called "wine and cheese" environmentalists, Malek-Wiley saw the march as a significant step toward forging a more inclusive environmental movement.[72] The march was intended to create new environmental activism in the state, and to a certain extent this happened. Many citizens did indeed join in the protest as it went through their communities. One Lafayette man, for example, commented as he entered New Orleans that he had "joined this march because I feel we all have a right to clean air and clean water. I'm going to march. I've marched all this way to say that." Environmental leaders such as Marylee Orr noted that the march gave a boost to their work. The official LEAN newspaper claimed that the protest "brought a new sense of power of individuals to an area blighted with chemical plant emissions."[73] OCAW activists certainly benefited from the march, as the visit to Geismar was widely reported on. Members of Local 4-620 walked much of the way, drawing some comment from the press; one journalist noted that "young, hippie-like Greenpeacers, complete with costumes, longish hair, shorts and sandals marched hand-in-hand with those tough looking, weathered, blue jeaned hard hatters."[74]

The march received extensive media coverage within Louisiana, both in the newspapers and on television. For its duration, it made the nightly news on all three major network television stations in both Baton Rouge and New Orleans. This level of coverage clearly helped to raise public awareness of environmental issues, as was graphically confirmed by a public opinion survey commissioned by the LCA. Carried out in December 1988, the survey was conducted in an eight-parish corridor along the Mississippi River. The survey contacted 450 registered voters in the eight parishes to determine their opinions on environmental issues in general, as well as their specific perceptions of the chemical industry. The survey concluded that the march, together with Greenpeace's activities, "did impact voters' level of concern about environmental problems," with 22.9 percent saying that their level of concern had increased, compared to only 1.7 percent who reported that it had decreased. The survey also revealed that most voters had a poor perception of the chemical industry, blaming pollution from chemical plants for "serious *health problems* and major *environmental problems.*" "Voters perceive state environmental laws and regulations and enforcement as unacceptably inadequate," it noted. Chemical companies were also "perceived as being *involuntarily* 'driven' to reduce pollution by environmental organizations and governmental agencies." Industry was clearly concerned by these findings, as their underlining of the most damning conclusions highlighted, and they seem to have influenced the LCA to place an increased emphasis on public relations in the 1990s.[75]

Shortly after the march, the union released its own video about the lockout. Produced by a documentary film company based in Washington, D.C., *Locked Out!* placed a strong emphasis on environmental issues. Peggy Hoffman, one of the St. Gabriel women worried about the miscarriage rate in her community, linked the chemical plants to residents' health problems. "To me, it is obvious that the industry has a problem or correlation with our health problems," she commented. AWARE leader Les Ann Kirkland expressed her concerns about the contamination of drinking water, while Amos Favorite described his personal experience of the area's high cancer rate. Other unidentified community members expressed their concerns that the BASF plant was not being run safely during the dispute. Several of the union's religious allies also agreed to appear, including Father George Lundy and Sister Fara Impastato. The union arranged several showings of the video in the midwestern communities where BASF was trying to build its new plant, successfully producing some public opposition to the German company in these towns. Several members of the public indeed claimed that they changed their mind about BASF after seeing *Locked Out!*[76]

The Christmas of 1988 was the fifth Christmas that BASF's maintenance workers had been away from their jobs. Throughout the dispute, Local 4-620's members had found Christmas to be an especially difficult time, as workers with children often lacked the money to buy them the presents that they wanted. On December 18, 1988, Ernie Rousselle dressed up as Santa and appeared outside the plant gates "to demand restitution." "For five years running the locked out workers and families have had their Christmases hijacked by BASF—We've been scrooged," he said. Santa demanded that BASF open the gates, adding that "Nobody—not even BASF—can be allowed to stop Christmas." Local 4-620's members were amused by Rousselle's stunt, commenting that he was "an awfully thin Santa." Despite the humor, however, by the start of 1989 BASF maintenance workers had each lost up to $140,000 in wages as a result of the dispute.[77]

8

Settlement

In December 1989, BASF finally agreed to recall all of the workers who had been locked out of the plant on June 15, 1984, a move that paved the way for the signing of a new contract. The agreement came after the union joined with environmental groups to launch an effective fight to stop the company from operating a new plant at the Geismar site until it had cleaned up contamination from under the plant. The ongoing campaign disrupted BASF's plans to expand its operations in Louisiana and contributed to its seeking a final settlement with the union.

Throughout 1989, the union concentrated on stopping BASF from operating a new plant to produce Glyoxal, which is a component of permanent press. The plant was built on the site of the former Basagran facility, and Miller argued that chemical contamination in the ground should have been cleared up before the plant was constructed. He asserted that BASF needed to excavate the site, a move that would involve dismantling its costly new facility. The campaign reflected Miller's ongoing efforts to disrupt BASF's operations, particularly its desire to expand the Geismar site. Lacking a strike weapon, the OCAW strategist argued that the union had to find alternative ways to exert pressure on the company. "We were trying to figure out how to stop production by using regulatory tools that were available," he asserted. ". . . We were driving to have them to have to tear the plant down and relocate it. That was our goal."[1]

Miller's efforts were fully backed by LEAN. In May 1989, for example, Marylee Orr wrote to the DEQ in support of the OCAW's position, noting that building over the site would prevent soil excavation. Like Miller, Orr worried that contamination from the production of the pesticide Basagran might migrate into residents' drinking water supply.[2] LEAN argued that the company had tried to conceal the contamination in order to ensure that the Glyoxal plant could be put into operation quickly. The environmental group argued that the site would "never be properly cleaned up" if the plant was allowed to operate, adding that this was not in the best interests of future generations who lived near the plant.[3] Answering BASF's accusations that the OCAW was "using" LEAN, Orr replied that her organization was "proud" to count Local 4-620 as a member. The two organizations did "work together on issues that are relevant to our concerns," she noted, but were otherwise independent.[4]

The revitalized DEQ took the concerns of OCAW and its environmental allies seriously and spent a great deal of time investigating contamination at the site. As the industry press acknowledged, the union and environmental groups were thus able to successfully hold up the new permit for BASF's Glyoxal plant.[5] In April 1989, a letter that Geismar site manager Bill Moran wrote to DEQ assistant secretary Maureen O'Neill revealed the effectiveness of these efforts. Moran explained that the company was under financial pressure to start its new plant: "Ms. O'Neill, we can not emphasize enough the urgency we face in reaching an expeditious resolution to this matter. At your request, the Air Quality Division is withholding approval of BASF's Consolidated Air Permit that BASF requires to begin operation of our new $12.5 million Glyoxal plant in August. Beginning Monday, May 1, the majority of the 28 new employees that will operate the Glyoxal plant are reporting for duty. This permit has been pending for two years and failure to obtain a prompt authorization to bring the Glyoxal plant on stream would cause severe economic hardship to BASF which could result in the layoff or reassignment of the new employees."[6]

Moran made clear that more was at stake than simply the Glyoxal plant, asserting that BASF was concerned that the failure to issue the permit might jeopardize its extensive expansion plans for the Geismar site. "Perhaps even more serious than the delay in bringing the Glyoxal plant on stream is the repercussions this could have on BASF's future expansions at the Geismar site," he noted. Moran explained that he had recently received authorization from BASF headquarters for "two major expansions" that would create more than five hundred jobs and cost $95 million. He reminded O'Neill of BASF's contribution to the state economy, noting that these jobs would be for "Louisiana residents." Further delays, he warned, would create "severe hardships for us and our employees."[7]

A great deal was indeed at stake for BASF. As the summer of 1989 began, the DEQ was withholding consolidated air permits for both the new Glyoxal unit and a construction variance for a $50 million PVP/NVP (polyvinyl) plant until the issue of contamination under the old Basagran plant was resolved. BASF lobbied hard to get these permits approved, visiting the state legislature on several occasions and securing a personal meeting with Governor Roemer. Roemer remembered that the company tried to press him to ensure that the permits were issued but that he refused to be swayed. The new governor was proud of his environmental record, and he did not intend to compromise it easily. Looking back on his administration after retiring from politics, he was keen to stress that he did not automatically grant permits to companies: "The normal way they're handled by governors is, 'What do you want? You got it.' This is, 'What do you want? We'll have to review it. It doesn't meet our standards. We appreciate you being in our state, but we don't appreciate you violating our environmental laws.'"[8]

This philosophy guided Roemer's meeting with BASF: "I remember the meeting well. . . . I remember the feeling of the meeting was their shock, and then their anger. . . . They had German representatives and local management at that meeting. Top executives, great shock, great disappointment, one of them told me, in Dr. Templet and Governor Roemer's reaction, but they obviously had not been keeping up with the position that we had been developing . . . which was that business is welcome, but they're not welcome to damage the environment. . . . They would have to prove their case before they got a permit. Can you imagine a worse meeting for them?" Although BASF denied it, Miller, Templet, and Roemer all saw the company's failure to persuade the governor as a turning point in the dispute. It was indeed shortly after this that BASF made a decision to try and secure a settlement with the union.[9]

As the summer of 1989 progressed, BASF continued to be frustrated in its efforts to run the Glyoxal plant. In July, the company received a further setback when O'Neill refused to approve a final corrective order that could have led to the permits on the new plants being issued. BASF attorney William R. D'Armand admitted that the company was "terribly disappointed" by this. "We would like to have an order," he acknowledged. Plant manager Bill Moran claimed that the company was not being treated fairly, arguing that other chemical companies were being allowed to clear up contamination using the pump-and-treat method that BASF proposed. Templet, however, asserted that pump and treat was no longer his "first choice," and he was trying to move the agency away from it.[10]

The DEQ also asserted that the company's soil tests were not detailed enough. BASF had tested the soil only for ethylene dichloride and monochlorobenzene, but the state now wanted the company to look for seven other chemicals. Templet presented his refusal to grant the permit as a long-term policy move, but he also acknowledged that the public pressure generated by the union and its environmental supporters also influenced his actions. These groups bombarded him with relevant information and ensured that he knew that every move by the DEQ was being closely watched. Representatives from the OCAW, LEAN, and the APRATP indeed attended every public meeting that the DEQ called to decide on the permits.[11]

Throughout the lockout, the union's environmental campaign worked because the chemical industry as a whole was very vulnerable to environmental pressure. Editorials and articles in *Chemical Week*, the voice of the industry, repeatedly highlighted the industry's concerns with improving its public image. Throughout the 1970s and 1980s, executives saw themselves as vulnerable to "monumental" environmental pressure and worried that they were losing the battle to convince the public that they were a safe and responsible industry.[12] In May 1984, for example, *Chemical Week* noted that the industry had "image problems," being linked in the public mind to toxic waste and pollution.[13] Executives had to respond by being much more proactive, establishing a better dialogue with the environmental movement, which was pictured as increasingly sophisticated and powerful. There was no room for "foot-dragging." Unlike the union movement, which they viewed as in decline, executives saw the environmental movement as a force of the future, and they worried that unions might resurrect themselves by make alliances with it.[14]

By the fifth year of the dispute, *Chemical Week* had become increasingly outspoken in its criticisms of BASF. The German chemical-maker was creating negative headlines for the industry, and the leading industry journal was not amused. In August 1989, *Chemical Week* called the dispute over the cleanup of the old Basagran site "BASF's Geismar debacle," adding that the company's new plant had been built on "questionable grounds." The article ran in an issue that included an editorial encouraging executives to address the needs of their increasingly "Green" constituents, an urging that seemed an implied criticism of BASF.[15]

BASF became increasingly isolated from the rest of the chemical industry as the dispute wore on. Industry executives disliked the negative publicity that they felt the dispute was causing, especially when it raised questions about plants' environmental record. Several other chemical companies were located next to BASF, including Rubicon and Borden, and Richard Donaldson remembered that the managers of these companies privately pressed local BASF executives to settle the dispute:

> We got more grief from the managements of other surrounding companies, Rubicon in particular, because we were drawing attention to them and they didn't like it. The union was going to the Department of Environmental Quality and saying, "Look, they're dumping stuff, they're spilling stuff, they're doing bad things, and so you need to do something about it." Once these guys looked up, they said, "Well, what's out there. Oh, you've got a pipeline that runs from BASF over to Rubicon." All this stuff was tied together. Borden was providing us with stuff. They were like, "Where does this stuff come from and how do you dispose of it, and who else is doing this?" So the people at Rubicon, they weren't real happy about that because they were discharging out with their permits. They were discharging hazardous materials on the ground, and they didn't really want anybody to take a hard look at that. So we got more grief out of them than we did from any of the [unions].

The union's environmental campaign, claimed Donaldson, was "extremely clever" and a "big, big nuisance." Les Story also remembered that management from other local companies put pressure on him to settle the dispute: "I think the other companies in the area, this is not a generalization, would have preferred not to have the chemical industry in the spotlight in Louisiana at this time of the decade. . . . When people started examining our records, our public records, they only had to turn the page and they could see the public records of the plant next door, and no one had ever looked at those before."[16]

Ed Steimel, the president of the LABI throughout the lockout, confirmed that other chemical companies, while publicly supportive of BASF, disliked the negative publicity that the dispute created and felt that the company should have settled it much sooner. "I don't think that other industry was lending them any particular support or hand in any way," he recalled. "I think people felt that this lockout just went on too long, and that there should have been some way to mediate it. . . . I did feel that this is one case where industry was maybe not working hard enough to settle the dispute. . . . I've heard others critical of BASF. . . . I haven't heard that kind of criticism, though, of other plants, so they were considered to be maybe partially at fault by other industry."[17]

The union's environmental campaign was also particularly effective because BASF wanted to expand the Geismar site. The company was deeply committed to the American market; between 1977 and 1987, for example, the North American portion of BASF's worldwide sales increased from 10 percent to 20 percent.[18] As a BASF executive commented in 1986, the company had "dedicated itself to expanding in the United States. . . . The United States is a tough market but it is also a market of great opportunity because it is the largest in the free world."[19] Just before the settlement of the dispute, company chairman Hans Albers himself described North America as a "top-ranking, strategic market for BASF."[20]

Within North America, Geismar was BASF's most profitable location and still offered the same advantages that had drawn Wyandotte to the state thirty years earlier, including excellent sources of raw materials, proximity to other manufactured chemicals, and good transportation links. In the late 1970s and early 1980s, Albers visited Geismar on several occasions and was reported to have "a soft spot" for the location.[21]

In the late 1980s, moreover, BASF's profits, like those of most chemical companies, were on the increase, and the company wanted to move ahead with its expansion plans. The company's annual reports showed that total sales, for example, increased steadily from $3.84 billion in 1986 to $5.4 billion in 1989.[22] "They knew they needed to get this labor dispute over with and get this company moving," asserted Richard Donaldson. The company's desire to expand Geismar is borne out by the fact that employment at the site doubled in the decade after the dispute ended. In 1990, BASF built a new $83.5 million plant to produce aniline and nitrobenzene. Aniline is a raw material for MDI (methylene diphenyl isocyanate), a high-demand chemical that was made at Geismar. MDI was used by BASF to make rigid foam insulation, as well as being combined with other chemicals to make synthetic fibers and high-resilience flexible foams. In 1990, Manfred Buller, the president of BASF's polymer division, noted that MDI sales had grown at "a 6 percent compounded rate" over the previous decade and were expected to continue to rise. The expanding Geismar facility was "a world-scale chemical plant," Buller told the *New Orleans Times-Picayune*. In the 1990s, BASF as a whole continued to grow, especially after the 1994 acquisition of English-based Boots Pharmaceuticals.[23]

Managers were also aware that local union members were unlikely to give ground. Since recalling the operators, Local 4-620's members had repeatedly refused to sign any contract that did not include jobs for the maintenance workers. Workers felt that they were fighting for this principle, rather than for economic gains. While the union obviously wanted to maintain a public face of unity, there is no sign that any union members wanted to give in. "We know we're right," commented Leslie Vann. "We'll fight them all the way to the cemetery, if that's what it takes." Asserting that the company's demands reflected its "arrogance," workers felt that they were fighting for their "dignity" and they were not prepared to give in easily.[24]

It would have been difficult for the union to have achieved the settlement without this complete unity from the rank and file. Many workers themselves commented on the militancy that the group had exhibited. "It was awesome," claimed Tommy Landaiche. "I've have never seen anything like it, never. I've never seen another union hold together like we held together, and I've seen a lot of groups on strikes and some on lockouts." In a December 1989 telegram to the local union, OCAW president Joseph M. Misbrener told Local 4-620's members that their "unity and tenacity" was "unparalleled." "You have been an example to us all and a major part of the history of the Oil Chemical and Atomic Workers International Union," he added. Ernie Rousselle explained that the BASF lockout was unusual because he never faced concerted pressure from union members to give in to the company and accept a weakened contract. He stressed that Local 4-620 members were a special group who were determined to make a principled stand against the company, despite the heavy costs involved: "Most people would endure a fight like this for a year or two years and then throw their hands up and say it ain't worth it. But these people decided not to do that, and I think that one comment which was made by one of our brother members at a union meeting I think emphasized that, and that is the comment, he said 'You can buy my work and you can pay benefits but one thing you can't buy is you can't buy my dignity, and I don't give a damn how much money you've got.'"[25]

The militancy of the BASF workers occurred in a region of the United States that was known to be strongly nonunion.[26] BASF workers were fond of saying that their determination had surprised the company, which had not expected southern workers to act in this way. "We were," acknowledged Henry Kramer, "a bit surprised by the intensity and staying power of the union." As such, the cultural stereotype of a docile southern worker motivated the group, making them determined to disprove it.[27]

The BASF workers actually drew on their culture to sustain them during the long dispute with the company. Many felt that BASF lacked respect for the area that it operated in and this motivated them to take a stand. Roy Fink, for example, felt that BASF was trying to "make fun of our culture. I've even done heard the comment, 'That ain't nothing but a bunch of dumb coon asses.' Well, we might be dumb, and our philosophy is, 'The first time you shit on me, it's your fault, but the second time you do it, it's my fault,' and you're not going to get the second chance." "Whoever gave them [BASF] their advice, to me just misread the people of southeast Louisiana," Fink concluded. Workers took pride in their "coon ass" culture and drew on it repeatedly to explain their militancy during the dispute, acknowledging that the company's foreign origins helped to motivate them. "Coon asses," argued Esnard Gremillion, had a strict sense of right and wrong, and they felt clearly that BASF was in the wrong: "They knew what was right and what was wrong, and what the company was trying to do was dead wrong in their eyes, and it was. They could have extended that contract and hammered something out in the next year or so. They could have done that, and probably saved money."[28]

Workers' culture also supported them in other ways. As was common in southern Louisiana, food served as a unifying bond for the locked-out workers. Almost every mass march or demonstration, including protests held out of state such as the "Lock-In," was concluded by a mass jambalaya cookout, prepared in the union's huge pot. These cookouts allowed workers to socialize with each other and with the guests that they had invited to their events. The jambalaya cooked by Local 4-620's members became so well known that the union even prepared the Louisiana dish for the entire OCAW convention on several occasions in the 1980s and 1990s. Even after the lockout, moreover, cooking out continued to play an important role in maintaining unity.[29]

Religion also played an important role in holding the workers together. In the post–World War II South, rates of churchgoing have been consistently higher than in the United States as a whole, leading many scholars to highlight the centrality of religion to southern culture. In many cases, particularly in southern textile mill villages, the church often played a role in discouraging southern workers from supporting unions.[30] This was not the case in the BASF lockout. Part of the reason for this was that many of Local 4-620's members, unlike most southern workers, were Roman Catholics, and the Catholic Church has generally been more supportive of workers' right to organize than Protestant churches. The vast majority of the locked-out workers attended mass regularly and were encouraged by the backing that the union received from Catholic leaders in the state.[31]

Smaller numbers of workers were Baptists and Methodists, and both of these churches also gave some support to the union. The fact that workers had been locked out, rather than on strike, helped them secure this backing, as they were able to portray themselves as victims of the company's actions. The importance of church support is confirmed by the company's efforts to persuade leading clerics not to support the union. Les Story, in particular, wrote to several church leaders to try and make them see the company's viewpoint, keeping separate files labeled "Catholics" and "Methodists" that documented his efforts. Unable to convince them, Story eventually became convinced that church leaders had been "brainwashed" by the union. OCAW leaders, for their part, recognized the value of church support, describing it as "extremely helpful."[32]

Other factors also helped to pave the way for a settlement. In February 1989, the NLRB denied the union's appeal against the board's refusal to issue a complaint against BASF. The board argued that BASF had not engaged in bad-faith bargaining and had thus been entitled to implement its final offer in October 1987 after the two sides had reached an impasse. The NLRB also accepted the company's argument that the subcontracting of maintenance work had been carried out "for purposes of efficiency" and not to break the union.[33] The OCAW was clearly disappointed by the decision. "Assurances contained in the New Deal era labor laws, that workers could not be discriminated against because they chose to act collectively, have become hollow," it noted in a statement.[34]

For union leaders, the decision epitomized the way that the NLRB in the 1980s became promanagement, reflecting the conservative appointments made by Presidents Reagan and Bush.[35] Over the course of its lengthy dispute with BASF, the union's experience with the NLRB was certainly mixed at best, and OCAW leaders were always aware that they could not rely on board charges to win the dispute for them. The board ruled against the union on several occasions, dismissing several of its claims that the company had bargained in bad faith. At the same time, however, the NLRB did reject BASF's attempts to stop the corporate campaign, a crucial decision that allowed the union to escalate its efforts, as well as providing it with a clear indication that the campaign was annoying BASF.

For the company, however, the NLRB's denial of the union's appeal was a validation and helped to pave the way for a settlement. Richard Donaldson remembered that managers felt that their decision to subcontract the maintenance work had been upheld and that it was now time to settle with the union. John Kirkman, who became BASF's vice-president of human resources in 1987, remembered that after the union's NLRB charges had been dismissed, the company "decided to 'turn down the temperature' of the rhetoric at the bargaining table." The company partly achieved this by removing Tom Budd as their chief negotiator. Described by Richard Donaldson as "the guy that the union just absolutely loved to hate," Budd was an abrasive attorney who had clashed repeatedly with Ernie Rousselle during negotiating sessions in 1988 and early 1989. Realizing that the personal chemistry between Budd and Rousselle was poor, Kirkman and another BASF executive met with Misbrener and Rousselle in Chicago during December 1989 to try and hammer out a deal. In return for Budd's removal, Rousselle agreed to step down as chief negotiator for the union. Negotiations were now headed by Kirkman and Bobby Schneider, an experienced BASF operator who had acted as the union's "inside" chairman since the production workers had been recalled. This change of personnel helped lead to the eventual contract.[36]

An agreement was reached because BASF gave ground on many of the issues that had helped to cause the dispute. Rather than a pay freeze, OCAW members received an immediate 2 percent wage hike, together with further increases of 3.5 percent in 1990 and 1991. The union also maintained its health coverage, with no increases in employee health care premiums for three years, as well as securing some pension improvements. Maintenance workers were given the opportunity to return as operators and were offered severance packages to compensate them for losing their maintenance jobs. Those who decided not to come back were paid $1,000 for every year they had worked for the company. The company also guaranteed workers' job security against plant shutdown or subcontracting through the seniority system. The union thus held onto its seniority clauses. "I can't think of any that they lost," admitted Richard Donaldson.[37] Union members were pleased with the settlement, which was ratified by an "ample margin."[38]

Company officials would have preferred to have secured the seniority changes that they had originally sought, but they recognized that they would have to give ground in order to bring the union back to work. The company's willingness to give way reflected the shift in approach that took place following the NLRB's denial of the union's appeal. BASF abandoned its attempts to secure productivity improvements and concentrated instead on making the changes that would bring the union back into the plant. As Richard Donaldson put it, "We changed our focus from trying to gain, to make any productivity gains to the old way of bargaining, and that is 'Okay guys, what is it going to take for you guys to come back to work, and we'll see whether we can afford it.'" The bottom line was that company officials knew that the plant could be operated profitably even with the old seniority system in place. The plant had indeed always been profitable prior to the lockout. Geismar's natural advantages—the size of the site and its access to the Mississippi River—continued to make it a very attractive location for BASF. By 1989, moreover, managers realized that operating the plant, even with the old seniority system, was preferable to having their business plans disrupted by a protracted battle with the union that showed no sign of ending.[39]

Reactions to the settlement varied. BASF officials were publicly tight-lipped about their reasons for coming to an agreement, although they did admit to being pleased that such a long dispute had finally ended. They also noted that their expansions would make it possible for them to recall all union workers to the plant. Both community leaders and industry groups expressed quiet relief that the dispute was over, especially as they felt that the union's campaign had tarnished the state's image. "I'm glad to see them go back to work," said Delores Shingleur, president of the Gonzales Chamber of Commerce. "I just hated to see [the dispute] go on that long." "Hopefully, we can all move forward," added Gonzales mayor Johnny Berthelot. "I'm thankful to the good Lord it's over."[40]

The union's allies, on the other hand, celebrated the signing of the contract. The *Toxic Times*, the newsletter of the Boston-based National Toxics Campaign, called the new contract "a monumental accomplishment against seemingly insurmountable odds." "The union struggle," it added, "spawned an activist coalition of workers, environmentalists, grassroots groups and other citizens who united to permanently change the face of Louisiana's social, economic, and political climate." Both workers and their environmental allies claimed that their campaign had exerted real pressure against the company and had led to tighter regulation of chemical companies. Willie Fontenot told the *New York Times* that the union had "raised the level of consciousness about the environment in Louisiana" and had helped to persuade the state to regulate industry more closely.[41]

The ending of the lockout was also widely celebrated by other allies that had supported the union. "What a heroic fight OCAW waged—for all of us!" wrote Ramsey Clark, the former U.S. attorney general who had represented those arrested during the lock-in. "And what a great bunch of individual heroes, too. People who put everything on the line—their safety, health, jobs, and life savings, family future—to outface enormous political and economic power and to protect us from more than Bhopal on the Bayou—from poisoning the planet." Bishop Stanley Ott added his praise. "I rejoice with you and all of the other people who have been affected by the labor dispute these past five years," he wrote Esnard Gremillion.[42] Several leaders of other OCAW locals sent messages of congratulations to their colleagues in Geismar, claiming that they provided an inspirational example for all union workers.[43] The AFL-CIO was similarly encouraged by the settlement, viewing it as "a hopeful augury for the decade of the 90s." Labor leaders sensed that the ending of such a lengthy dispute was significant. "The BASF lockout will go down as a landmark in labor history," noted Carl Crowe, an official of the Louisiana AFL-CIO.[44]

OCAW leaders also saw the settlement as a significant victory. They had conducted the lengthy campaign to make a stand against concessionary bargaining, and they felt that their position had been validated. Robert Wages commented that the union had "sent a powerful message to anyone who would try to deprive our members of their dignity, their livelihoods and their union. And that message is that we have the will and the capacity to defend ourselves even under the most adverse of circumstances."[45] Locked-out workers themselves proclaimed a victory against the odds. "We took on a hostile company, a hostile government, and we won," exclaimed Leslie Vann. "We're back in the plant."[46] At a party held to celebrate the ending of the dispute, locked-out workers and their community allies posed for the cameras holding signs that proclaimed "Victory!"

This public proclamation of victory concealed the fact that the lengthy dispute had cost both sides a great deal. Since June 1984, the union had spent more than $3.5 million on benefits for the locked-out workers and on the campaign.[47] The OCAW also estimated that its members had lost over $50 million in wages and benefits as a result of the dispute, and more than fifty homes had been repossessed.[48] BASF officials refused to reveal what the lockout had cost them, even in retrospective interviews. The sum must have been considerable, however, especially as those involved in the dispute all stressed the large amount of resources they had to devote to answering the union's allegations. One salaried worker at Geismar, for example, claimed that the union's campaign "did cause the company a lot of headaches and money."[49] In July 1988, one union estimate claimed that the lockout had cost the company over $19 million, including over $1

23. Local 4-620's Victory Party, January 1990. (Courtesy PACE International Union)

million on legal costs. The exact figure is very difficult to pin down, however, especially as much of the union's effort was directed at depriving BASF of the potential profits it would have made from new and expanded plants.[50]

In some ways, the BASF lockout was not quite the convincing victory that the union publicly claimed. The union lost the maintenance jobs, a clear setback, especially as many union leaders had traditionally come from that department. The BASF "victory" was also a defensive one, as workers had fought to uphold the status quo, returning to work with a contract similar to the one they had prior to the dispute. Against a company that had sought major concessions, however, just holding the line seemed like a victory. Since 1981, few local unions have been as successful at resisting concessionary bargaining as OCAW Local 4-620 was. Some workers, afraid of striking, have accepted major concessions and returned to work with a weak contract. When workers did walk out, many companies hired permanent replacements, a move that often led to the decertification of the union involved, as replaced strikers lost their rights to vote in any ensuing representation election.[51] In the 1980s and 1990s, countless local unions were decertified in this way. In this context, the Geismar union had secured a real achievement in being able to resist the company's concessions and secure a new contract. "In the eighties," commented Richard Miller, "the victory was avoiding elimination."[52]

The union had, moreover, achieved what it had always wanted—to provide all its members a chance to return to the plant. Although maintenance workers lost their jobs, they retrained into new positions that were often as well-paid as those they had held before the lockout. Many workers who went through the dispute felt that they had established an important point of principle. "I considered it a victory for us, to have went [sic] as long as we did and to get it settled," reflected Bobby Schneider. ". . . We did retain our seniority, which they were after, and we made improvements in it. We did get those hundred and ten people that were out there that they said there was no way in hell they'd ever come back out to that plant. They at least got the choice of going back to work." "I feel like we proved that we were willing to go to any lengths to be treated with respect and dignity," added Roy Fink. ". . . What we did win was the fact that they know we're not scared to take and lock horns with them, if we feel like we're right."[53]

Although the majority of workers wanted to return to BASF, 53 of the original 370 either resigned or took an enhanced retirement package. Many older workers, in particular, chose not to go back after the lengthy dispute. Among them was Esnard Gremillion, who took early retirement in January 1990.[54] BASF was able to offer jobs to all those who did choose to return. "One of the things that helped us so much in bringing the labor force back into work at Geismar was the expansions," explained Richard Donaldson. "If the site had stayed the way it had been in 1984, I don't know what we would have done with all those people." Most of the former maintenance workers were given jobs in BASF's new aniline plant, and they adjusted well to production work.[55]

Even though the maintenance jobs had been lost, many former maintenance workers felt great satisfaction in returning to BASF because they felt that the company had wanted to "get rid" of them. Roger Arnold, for example, chose to return to BASF in a lower-paying job because he "wanted to show that they were going to bring me back. . . . I said it during the lockout and negotiations, that I'm coming back, and that's the way it was." Like many workers who went through the lockout, however, Arnold was also acutely aware of the costs that workers incurred. "In the eyes of getting the people back to work and having a stable job and all that, yeah it was a victory," he reflected. "It was a victory in straightening up the environmental issues of the plant, and all the plants involved, getting involved with the communities and all, as far as environmental issues are concerned, the safety issues are concerned, is a big, big, [plus], but losses of what happened during the lockout, such as not being able to go fishing and hunting with your son or your daughter, that's something that you'll never get back. You can't get it back, there's no way, and that will go with me, I guess for the rest of my life. Do I resent that? Yes, I do. I'd be crazy to say, 'No I don't,' because I do, I resent it. It hurts, don't ever think it don't."[56]

Putsy Braud, a former maintenance worker who had been at the plant since it opened, also had mixed feelings as he returned to work:

There was a little satisfaction to go back in and there was a lot of hesitation too. I mean there was a lot of people in there that had took our jobs from us and we had to go back in there and work with them, and I didn't like that, but a little bit of satisfaction knowing that we had the opportunity to go back. I know they didn't want us in there and they still don't want us in there, but there's not too much they can do about it. They found that out. They tried their damnedest to get rid of us but they just couldn't do it, and people pulled together long enough. That's what union is all about; if you stick together long enough, then you can win it, but it took its toll on a lot of people. A lot of people committed suicide, some of them lost their families and homes and everything. A lot of them couldn't find real good jobs. There just wasn't that many out there. In the eighties there was some bad times, economically it was bad. There wasn't too many jobs, and a lot of people just couldn't handle it.[57]

No More Trade-Offs

Although the contract formally brought the dispute to an end, relations between the union and company had been permanently transformed by the lengthy conflict. For many union members, one of the most pleasing aspects of the settlement was that they had not made any pledge to stop their environmental work. Throughout the 1990s, in fact, Local 4-620 continued to work with environmental groups and funded its own organization, the Louisiana Labor-Neighbor Project (LLNP), to coordinate its efforts.[1]

Workers themselves acknowledged that they had been initially attracted to working with environmental groups primarily as a way of exerting leverage against the company during the lockout. Over time, however, their attitudes changed, and they began to see the value of raising environmental issues even when there was no labor dispute with the company. Workers thus described how their they received a tremendous education about environmental issues in the course of the dispute. "Of course we had a vested interest because a little pressure on them would help our cause to get back to work," acknowledged Carey Hawkins. "So I'm sure at first a lot of people felt that that was our main interest, but then as time went on, I think that it was found that this is more important even than our job, that someone needs to watch and ask questions. Are environmental regulations being followed? So it was sort of an awakening I guess to a lot of people, but hey look, you know, it's more than just a job, this is important."[2]

Bobby Schneider also described the transformation of workers' attitudes to the environment as an "awakening."[3] "Five years ago," he commented in 1990, "I felt if you wanted your job and you wanted to live in this state, you were just going to have to put up with it. I certainly don't feel that way today. There are ways to make these plants safe—to cut down on wastes, to recycle wastes, and to find proper ways to get rid of the waste that's left. A company told me that with all the trouble they were having they might have to build in Texas. I told him the plant gate was about a mile from here—don't let it hit you in the ass on the way out."[4]

All the workers who went through the lockout described a similar transformation. Putsy Braud, who had worked at the plant since it had opened in 1958, remembered that the lockout permanently changed his environmental views: "You've got to realize that everybody has to do their part and keep this planet as clean as possible for generations to come. . . . It made the company, they're more aware of it now, the things we used to dump on the ground. You had a bucket or something, you'd just throw it on the ground. Now, you don't do that now. There's a place to put it and recycle it, get it back into the system or put it in a waste tank, or whatever, haul it off to be properly disposed." This transformation also affected workers' personal lifestyles, as many claimed that they had started to recycle at home and had become more aware of the need to conserve resources since the lockout.[5]

Many union members used conservationist arguments to explain their support of stricter environmental regulations. Workers who liked to go hunting and fishing argued that continued pollution threatened to destroy their enjoyment of these activities. "I love going out fishing," noted Roger Arnold. "As a matter of fact, this past weekend me and the wife went to her uncle's place over there, and we set out lines to try to catch some catfish. We didn't catch nothing but that's what we like to do, and we don't want the water contaminated so bad that you can't do that. So we've got to clean the waterways up, and that's what we done." Like other workers, Arnold especially wanted to ensure that the environment was cleaned up so that future generations would be able to enjoy hunting and fishing. He felt that his views about the environment had been permanently changed by the lockout, asserting that workers' relationship with BASF could "never go back to like it used to be."[6]

At the end of the lockout, the BASF workers were no longer afraid of losing their jobs if they spoke out in favor of environmental issues. "They keep threatening us with the loss of our jobs," noted John Daigle in 1990. "But we don't want those kinds of jobs anymore. We want clean jobs."[7] Many asserted that the plants, including their own, should be shut down if they were not as clean and safe as they could be. Workers were no longer willing to see pollution as a necessary trade-off for having well-paying jobs. "I don't believe we need to pollute everything there is in order to have a job," asserted Gladys Harvey. Rather than being representatives of the chemical industry, workers now viewed themselves as Louisiana residents first and foremost. They were willing to envisage a future without the chemical plants, arguing that the state had been much more beautiful before the plants arrived. "What they think is that this is such a great asset to us because we can make money, but you're killing us," asserted Frank Smith. "I stand in my backyard sometimes and smell chlorine."[8]

For a few workers, environmental work became their life because of the dispute. During the lockout, maintenance worker Darryl Stevens worked with LEAN and the APRATP to collect data that monitored air, water, and soil pollution. Stevens' wife, Ramona, also began to work for LEAN, and the pair decided to commit themselves to the environmental movement once the lockout was over. "We're an unusual couple," he admitted in 1990. "Normally, on Saturday nights, couples go to movies and out to restaurants. We go to the chemical plants and 'get polluted.'" While Ramona Stevens carried on working for LEAN after the lockout, her husband set up his own environmental consulting business.[9]

The vast majority of workers were, however, happy to return to the plant after the dispute, arguing that this right to work at BASF was what they had been fighting for. Workers also justified returning to the chemical plant by arguing that only a union workforce could ensure that it was operated safely. Union workers took credit for the reduced emissions that BASF publicized in the early 1990s, asserting that pressure from Local 4-620 had forced the company to improve its environmental record.[10]

After the lockout, the OCAW workers who returned to the plant began to monitor BASF's environmental practices with a new vigilance. In 1990, for example, the union's complaint to OSHA led to a $320 penalty against BASF for operating an improperly repaired cracked absorber tower.[11] In November 1991, meanwhile, the union claimed that BASF had committed eight OSHA violations after four OCAW members were exposed to toxic chemical vapors without being given adequate respirators.[12] After filing many complaints like this, the union soon gained a reputation as an effective whistle-blower against environmental malpractice by chemical companies along the Mississippi River. "We even have other plants call us," noted Roy Fink, "right there on the corner, and tell us stuff, 'Look, hey, can y'all get such and such checked?' or 'They doing such and such, or they doing this or they doing that,' and we've done it. We've done it and we'll continue to do it. . . . They know through being associated or being kin with people like us, union people, to say, 'Hey, we'll call [Local] 4-620. I bet you they'll do something about it,' and I mean we don't just get a phone call and just act out of frustration or just act off of that; we check it out to make sure that it's genuine and that we really have something. We don't hesitate."[13]

With their attitudes to the environment transformed, Local 4-620 members were also keen to continue their work with local residents. Richard Miller stayed in Louisiana for two years after the ending of the dispute and was particularly anxious to see that the union continue its coalitions with the community. "We've got to carry forward the momentum around these organizing victories—around toxics, taxes and other social justice issues," he declared. "What we've discovered is that our strength here comes more from coalitions with the community rather than from the labor movement. We're trying to institutionalize these gains so that this local union will never again be left naked to a union-busting company." As such, Miller felt that a permanent alliance with the community would provide the union with a weapon that it could use in any future fight with BASF.[14]

In a series of meetings held shortly after the end of the lockout, the local union agreed to try and set up a permanent environmental program that would be funded by members' dues. Anxious to make the project as broad as possible, Miller was keen to involve the Vulcan workers, who also belonged to Local 4-620. Many Vulcan workers were friends with BASF workers, while others were related. Encouraged by Miller, BASF workers persuaded many of their Vulcan counterparts of the need to work with environmentalists. Duke King, a union leader in the Vulcan plant, recalled that although some workers initially feared that the alliance with environmentalists could cost them their jobs, the union educated many of these workers that a clean environment did not necessarily lead to job losses. In the 1990s, Vulcan workers' own experience gave some support to this argument. The Vulcan plant made several ozone-depleting chlorofluorocarbons (CFCs), chemicals that were gradually phased out by the ozone preservation provisions of the 1990 Clean Air Act. Some workers initially worried that this might lead to the plant's closure. In reality, however, the plant simply switched to making other chemicals, and no workers lost their jobs as a result.[15]

By observing the lockout and talking with BASF workers, the environmental views of many Vulcan workers had also been permanently transformed. "The Vulcan people had stayed pretty abreast of it," recalled King. "They understood the help that the environmental people had given to the union. . . . The whole union was affected from an environmental standpoint. We was all educated about the value of working with environmentalists. We all profited from the standpoint of being educated about the environment." Like many BASF workers, Vulcan workers had become genuinely more concerned about the environment, although they also recognized that working with environmentalists gave the union a weapon against the company, protecting it against concessionary bargaining. Vulcan workers had seen how the BASF workers' tactics had helped them to resist the company's concessionary demands. Aware that they could also face similar demands in the future, many Vulcan workers saw the need to reach out to the community as their BASF counterparts had done.[16]

On March 8, 1990, a special union meeting was held to vote on a one-year dues increase of $5 per month to fund the program. Although many BASF workers were still in debt after the dispute, the proposal passed by a 78.7 percent margin. While many Vulcan workers supported the project, Miller admitted that some had been impossible to convince. Nevertheless, he was pleased with the solid backing that the union had given to the program. The LLNP was born.[17]

The LLNP aimed to work with community groups and workers in order to achieve what it called "social, economic, and environmental justice for low-income and working families in Louisiana."[18] It aimed to "institutionalize some of the organizing and environmental policy gains" of the lockout by creating a "long term labor-community coalition."[19] The idea of the LLNP was strongly supported by OCAW leaders and local union members, who both were anxious to prove that the union would not abandon its environmental allies because the dispute had been settled.[20]

Several other environmental initiatives emerged out of the lockout. At the end of 1989, the union helped launch Louisiana Workers against Toxic Hazards (LA WATCH), which was designed to organize and train workers around workplace-safety issues. LA WATCH argued that federal health and safety regulations were not adequately enforced. In 1989, for example, there were only seventeen OSHA inspectors to protect over 1.2 million Louisiana workers. Operating out of Local 4-620's headquarters, LA WATCH set up a hot line for workers who were concerned about environmental malpractice in the workplace. The organization was supported by a coalition of unions and environmentalists, including the Aluminum Workers Union, the Baton Rouge AFL-CIO, LEAN, the United Steelworkers Union, and the Service Employees International Union

(SEIU). LA WATCH struggled to secure adequate staffing, however, and its work was quickly relegated to that of the LLNP. In 1991, the local union voted to end the program and concentrate instead on the broader environmental work that the LLNP was engaged in. "The environmental thing just kind of took over," remembered Duke King.[21]

Another coalition operating out of Local 4-620's offices was the Louisiana Coalition/Citizens for Tax Justice (LCTJ), which was modeled on the labor-backed national Citizens for Tax Justice. The LCTJ grew directly out of the lockout, particularly Miller's work to stop BASF from operating its new Glyoxal plant. During this campaign, the union had discovered that ten-year property tax exemptions were granted by the state to companies that constructed new plants, even if they had not complied with environmental law. After trying to stop BASF from receiving its tax exemption for the Glyoxal plant, the union decided to continue working on this issue, securing funding from both the Ford Foundation and an anonymous donor.[22]

The tax exemption campaign mobilized support from citizens who felt that Louisiana's public services, particularly its low-ranking public schools, were suffering because of tax breaks given to industry. By October 1992 the LCTJ was supported by more than thirty-three local and statewide groups, including grassroots environmental organizations such as the Calcasieu League for Environmental Action Now (CLEAN), the APRATP, and Delta Greens. Union support, meanwhile, came from the Carpenters union, the United Automobile Workers (UAW), the SEIU, and the OCAW.[23] In the summer of 1992, the LCTJ lobbied intensively for tax reform, its Tax Justice Caravan crisscrossing the state to spread its message. The group also protested outside the Board of Commerce and Industry in Baton Rouge, a key body in deciding the award of tax breaks. LCTJ activists argued that the board was dominated by industry and should include a more diverse sample of Louisiana citizens.[24]

Headed by Zack Nauth, a former reporter with the New Orleans Times-Picayune, the LCTJ carried out a major study of the state's tax system that attacked "corporate tax giveaways."[25] Published in 1992, The Great Louisiana Tax Giveaway argued that between 1980 and 1989 the state had forfeited more than $2.5 billion in revenue because of the property tax exemption.[26] The tax relief was justified as a way of creating jobs, but the LCTJ argued that 75 percent of it created no new permanent jobs. Over 94 percent of the money indeed went to existing large corporations. The 1980s, it asserted, had been a "decade of corporate welfare."[27]

Industry, however, insisted that it was already the biggest taxpayers in the state and fought hard to stop tax reform. "The people who allege we're not paying taxes don't know what they're talking about," asserted Dan Borne. Despite the LCTJ's lobbying efforts, the status quo was also endorsed by the majority of state legislators. Faced with this opposition, the LCTJ was unable to secure a reform of the state's tax laws and was a dormant organization by the late 1990s. Despite this, the campaign had increased many citizens' awareness of the tax breaks awarded to industry by the state.[28]

In the early 1990s, Local 4-620 also continued its fight to clean up the former Basagran site. In May 1991, Richard Miller joined with Amos Favorite and Thomas Estabrook of the National Toxics Campaign to urge the DEQ to require BASF to excavate the site. The three activists rejected the company's argument that the contaminated groundwater under the site would never be used by residents. "We feel," they noted, "that it is not BASF's right to dump toxic chemicals in a public resource, regardless of whether they assume that the groundwater will never be used for drinking water."[29]

In September 1991, the company submitted a cleanup plan to the DEQ that for the first time accepted the need for some excavation, a move that the union took credit for.[30] After Edwin Edwards was reelected, however, the union complained that the DEQ became less responsive to its campaign. In 1992, the environmental agency amended an earlier order that had required the company to excavate two areas, insisting that this was no longer necessary. Without excavation, the union complained that the site would take more than two hundred years to clean up.[31] After Miller left Louisiana, his work was carried on by Dan Nicolai, who was hired as Local 4-620's "environmental coordinator." Nicolai tried to keep the pressure on BASF, telling the DEQ that the company's plans to pump chemical contamination out of the ground were inadequate. He cited a study carried out by Steve Amter, a groundwater hydrologist hired by the union, that showed that BASF's proposal might work to remove chemicals that mixed easily with water but was unable to eliminate patches of pure chemical believed to be underneath the plant. BASF, however, denied that the contamination was this extensive and pressed ahead with its plan. In the late 1990s, BASF was still treating contamination at the site using its pump-and-treat plan and had not carried out any excavation.[32]

Local 4-620 also played an important role in encouraging the residents of Geismar to fight for a clean drinking water supply. In 1991, the union, in association with the APRATP and the National Toxics Campaign, conducted a survey of 940 households in Geismar and the neighboring community of Dutchtown. The survey revealed that 82 percent of residents were not happy with the quality of their drinking water, which was drawn from private wells susceptible to chemical contamination. Over 60 percent of those surveyed were already buying bottled water. "Sometimes the water tastes and smells like bleach and I get headaches from the smell," complained one resident. "I don't trust the water and that is why I buy it instead."[33] The OCAW indeed claimed that water quality was threatened by poor sewage treatment, a nearby hazardous waste site, the underground injection of waste, and leachate from radioactive phospho-gypsum waste piles. Eight out of ten respondents wanted a municipal water and sewer system.[34]

After recruiting local activists who wanted to work on the issue, the LLNP formed a biracial steering committee that organized local citizens into the Geismar/Dutchtown Residents for Clean Water and Air. On August 28, 1991, the group's first meeting drew 175 people, including several OCAW members, and secured backing from the local volunteer fire department and several churches. After being addressed by Richard Miller and Zack Nauth, the group decided to push for a municipal water system. They met with the parish police jury and lobbied for a new piped water system that would bring in uncontaminated water from Baton Rouge. In March 1992, Amos Favorite became part of a board appointed by the police jury to plan the new system. Although the group tried to gain industry support, industry denied that the residents' drinking water was contaminated and argued that it was working hard to reduce its emissions. Despite this lack of support, the residents were ultimately successful in securing the new water supply. They set up a community Waterworks District and Water Board and gained parish funding for an engineering feasibility study. After the study showed that it was feasible, residents were able to secure a clean water supply from Baton Rouge Water Company.[35]

The setting up of the municipal water system was a major achievement and solidified the relationship between Local 4-620 and the Geismar residents. "That was a very, very successful project," noted Willie Fontenot, "and it brought something to the community, and it was organized and supported by the Oil, Chemical, and Atomic Workers' Union and the Labor-Neighbor Project. Having this public water system set up, having it become successful, gave the local community some real power and a real resource that they didn't have before."[36]

The union continued to fight for the rights of residents in other ways. In September 1993, local union members joined with residents of Geismar and Gonzales and traveled to the DEQ to demand an end to what they called "dangerous emissions" coming from the Rubicon plant in Geismar. The activists argued before a DEQ hearing that emissions of benzene, a suspected carcinogen, were making area residents sick and that it was time for the company to improve its environmental record.[37] In 1994, meanwhile, a group of Geismar and Gonzales residents and OCAW representatives met with management of Vulcan Chemical to express their concerns about releases of hazardous chemicals from the plant.[38] In 1995, the LLNP also worked with the Geismar/Dutchtown Residents for Clean Air and Water and convinced the DEQ to install a twenty-four-hour air monitor in their community.[39]

In April 1994, the LLNP and OCAW helped organize a coalition of community groups, including the Geismar/Dutchtown Residents, Concerned Citizens of Gonzales, and the Plumbers Union Local 198, to fight successfully for a parish ordinance that ensured better employment opportunities in the petrochemical plants for parish residents. Ascension Parish became the first parish that required companies to report how many of their employees (including contractors) had been parish residents for at least one year. Despite opposition from the Ascension Parish Chamber of Commerce, community and union members helped to persuade the parish council to adopt the ordinance. Since the ordinance, the plants have hired a greater percentage of workers from Ascension Parish; at BASF, for example, around 40 percent of workers were parish residents by August 2000, a clear increase compared to the early 1980s. More aware of residents' needs, some workers who went through the lockout have also moved from Baton Rouge to Ascension Parish, while others guiltily admit to still living outside the parish.[40]

The LLNP has been the most successful and most enduring of several projects set up by the union after the lockout. The organization continued to exist more than a decade after the end of the dispute and was still funded by the union dues of Local 4-620's members, although it had also been successful at securing additional outside funding. Throughout the 1990s, the LLNP was involved in a wide variety of projects, placing a strong emphasis on developing grassroots leadership and improving the quality of life in local communities. In the early 1990s, for example, the group worked with local environmental groups to fight discharges of sewage from a state prison in Iberville Parish into recreational and wildlife areas, a battle that resulted in the construction of a $1.1 million sewage treatment facility. It also worked with community groups to secure new equipment and repairs to parks in local African-American communities. In the local community of Darrow, the LLNP helped organize citizens' groups to knock down or clean up abandoned houses, while in the river-side town of White Castle it fought for more funds to carry out structural improvements to low-income families' homes. It also helped to organize candidate forums, allowing local people the chance to present their concerns to those who were running for local political office.[41]

One of the LLNP's most notable achievements was its fight, together with a broad coalition of local and national environmental groups, to prevent a $700 million polyvinyl chloride plant from locating in a poor, predominantly African-American community in St. James Parish. In August 1996, Shintech, part of the Japanese company Shin-Etsu, gave public notice of its plans to build the plant, having already secured the support of Governor Mike Foster and St. James Parish president Dale Hymel. Many residents organized against the proposal, however, setting up the St. James Citizens for Jobs and the Environment to coordinate their efforts. They were assisted by activists from groups that included Greenpeace, the Southern Christian Leadership Conference, LEAN, and the United Church of Christ Commission for Racial Justice.[42]

In March 1998, the LLNP joined these groups to cosponsor public hearings in St. James Parish. In a subsequent report, the company's opponents claimed that "an overwhelming presence of citizens" had attended the hearings, providing "compelling statements about the existing environmental degradation and health problems that would be worsened by Shintech's pollution." Shintech's opponents concluded that siting the plant in St. James Parish represented "a classic case of environmental and economic injustice."[43]

This battle against Shintech Corporation received a great deal of press attention and became a national test case for the EPA's environmental justice policy. Between 1996 and 1998, the staff of the LLNP spent a great deal of time working with residents and encouraging them to oppose the Japanese company. In addition to carrying out research and putting residents in touch with Greenpeace, they also worked to educate local people about "job blackmail," arguing that local communities were unlikely to secure the substantial economic benefits from the plant that the company promised. In a move that many believed was a reflection of the company's desire to avoid further negative publicity, Shintech eventually abandoned its plans to build in St. James Parish, although it did construct a smaller plant in Plaquemine. The Shintech fight helped the LLNP to become well-known in Louisiana, and by the summer of 2000 the group had been successful at securing enough outside funding to allow it to employ four full-time staff.[44]

The lockout also encouraged some local residents to become active in the environmental movement. In White Castle, located immediately across the Mississippi River from Geismar, Albertha Hasten watched the dispute and felt inspired to start fighting the pollution problems that affected her community. In the late 1980s, White Castle was predominantly African-American and poor, with most residents eking out a living by farming sugarcane. It was also surrounded by the sprawling chemical plants in St. Gabriel, Geismar, and Donaldsonville. Hasten, an African-American mother with two children, had never been active in the environmental movement until the lockout caught her attention: "I heard about the lockout and amazingly, it was right across the river from me, two miles, and I wanted to know. So I asked questions. I called and LEAN gave me some information on it . . . and I met Mr. Amos [Favorite], Leslie Vann at LEAN because they was on their board and the stories that I heard and the things that they said, inspired me, to look at not just my situation but to look at the whole picture and what is going on." Hasten was especially impressed by the way that the union carried on its environmental work after the dispute had ended: "I figured that was all they was concerned about, their wages, and they was concerned about having better facilities and a better job, but after the lockout was over, they organized communities, to build a bridge, a connection with other communities and have a strategy plan on what you need to do, how they need to do it, and when you need to do it, because getting organized is so important, and they didn't want it to stop. They didn't want their work to be in vain. So they paid membership dues . . . and they're still out here doing work."[45]

In 1990, Hasten formed Concerned Citizens of Iberville Parish, a local environmental group that voiced residents' concerns about industrial pollution. A decade later, she was a LEAN board member and was well-known for her activism across the state. Hasten herself claimed that the lockout had inspired many local people to become involved with the environmental movement. "We wouldn't never have had neighborhood organizations if it had not have been for the lockout, because it was a chance to see that something can be done," she asserted.[46]

LEAN continued to be active throughout the 1990s, and Local 4-620 remained one of its board members. When the statewide organization was founded in 1986, it was composed of only a handful of local environmental groups, but by the end of 2000 its member groups numbered eighty-five.[47] LEAN had become a well-known authority on environmental matters; in September 2000, even the LCA's Dan Borne called it "a very active voice, a very respected voice."[48] LEAN has been greatly helped by the 1987 Emergency Planning and Community Right to Know Act, which requires manufacturers to annually report the total amount of 330 toxic chemicals that they have released into the air, water, and land. The act was prompted by the Bhopal disaster and passed despite opposition from industry and the Republican White House. In the decade after the act's passage, LEAN used it to acquire data on pollution and publicize the record of companies who were the biggest polluters.[49]

Under this pressure, Louisiana's toxic emission levels did fall in the 1990s, dropping by more than 50 percent in one three-year period in the middle part of the decade. Although industry disputed LEAN's claims to be responsible for this drop, it accepted that the right-to-know law had played an important role in pushing companies to cut emissions. Local 4-620's members also claimed that their increased activism was partly responsible for declining emission levels, particularly at BASF. Between 1987 and 1995, the company achieved a 97 percent reduction in overall emissions. Rather than relying on underground injection wells to dispose of waste, a method that the union had fought during the lockout, the company had pioneered new environmental engineering procedures to find markets for waste that was previously deep-well injected. In the early 1990s, the company also constructed the only pure-oxygen wastewater treatment plant in the country. In 1994, the DEQ recognized the environmental progress that BASF had made by awarding the company its annual environmental leadership award. "BASF-Wyandotte," commented Amos Favorite, "was the worst company we had out there one time, but they got to be the best."[50]

Union workers' willingness to file complaints against BASF contributed to the poor relations that existed between them and the company for many years after the dispute. In April 1990, for example, Richard Miller spoke of abundant "morale problems" among Local 4-620 members, many of whom felt that managers did not want union workers in the plant and were consequently favoring the contractors.[51] Managers claimed that they tried to overcome these suspicions and complained that many union members had an "attitude problem." Over time, relations between the two sides did improve, yet most union members who had been through the lockout felt a lasting distrust of the company.[52]

Although BASF was keen to assert that relations with its workers eventually returned to "normal," workers who had been through the dispute never forgot it and lived in continual expectation of further conflict. "I'm always looking for where their next knife is coming from," explained Gladys Harvey. "They won't get me again next time. I've scrounged the years that we've been back and they can lock me out tomorrow, no problem." Other workers were even more forthright. "I'll never trust them bastards if they're alive in another five hundred years and I am too," asserted Bobby Schneider. "I'll never trust anything that comes out of their mouth. Players have changed, the game is the same. . . . If their lips move, there's a good chance they're lying, a good chance."[53]

Neither Harvey or Schneider felt the same pride in working for BASF after the lockout. "I no longer flash my badge and say, 'I work at BASF,'" explained Harvey. "I say, 'I work at a chemical plant down on the river.' I don't have any pride as far as their name or anything like that. I used to be pretty proud of my job. Now I just say, 'I'm an operator.' It's not what it was when I first started at all. I don't feel any loyalty to them. It's still there inside. It's tough. I still do my job. People do their jobs not for their company, they do their jobs for themselves." "I don't want nothing from them but a paycheck," added Schneider. "I don't care to go out and eat supper with them. I don't care to socialize with them. I go out there and do my job with them and that's all I care to see. The ones of us that went through it are the same, and that has had a bleed-over effect on some of the other ones."[54]

The two sides were able to agree to new contracts in 1992, 1995, and 1998 without further labor disputes. Workers were generally satisfied with the contracts, and BASF had clearly abandoned its earlier efforts to seek wholesale concessions from the union. Some workers were, however, dissatisfied with the company's failure to offer them a 401K plan, a common part of the benefits package available to nonunion chemical workers.[55]

The local union has remained strong throughout the 1990s. Following the signing of the 1989 contract, many contract workers who had been hired during the lockout stayed on in the plant and secured permanent positions. Initially, these workers were not welcomed into the union. As their number grew, however, local union leaders quickly realized that the presence of increasing numbers of nonunion workers was weakening the local. In March 1990, they proposed that former contractors be included in the union. After a divisive and emotional union meeting, Local 4-620 voted to put the dispute behind it and extend the hand of friendship to the former contractors.[56]

For most workers, this decision was a very difficult one to make, but they realized that they needed to act in the best interests of the union as a whole. "We had a long discussion on the floor about that before we voted on that, absolutely," recalled Bobby Schneider. "There was lot of hard feelings there. 'I don't want him belonging to my union when he had worked my job.' It was a long night that night and we covered all the points I thought well and it just came down to it to where we just had to make the decision, if we were going to survive, that's the way I looked at it, because I could see in the future the way things were heading. The numbers would have been there to get a decert vote and maybe wind up without a union at all. So it was a heated debate and there was people who voiced their opinion very strongly against it, but the vote carried to extend an invitation to them, which we did." After the meeting, union members began an organizing campaign, visiting all workers who were not in the union and encouraging them to join. Through their determined efforts, they gradually reduced the number of nonunion workers. In August 2000, just two workers out of nearly five hundred were not in the union. Although Louisiana was a right-to-work state, Local 4-620 had managed to secure almost 100 percent membership levels.[57]

The union's environmental initiatives also meant that BASF was pushed into its own efforts to reach out to the community. Throughout the 1990s, in fact, the company made an increased effort to address residents' concerns, although it obviously did not portray these moves simply as a response to the union. In 1992, for example, BASF was the driving force behind the setting up of a telephone number that local residents could call if they had questions about odors, noise, or other problems relating to the Geismar chemical plants.[58] A year later, the company began to publish its own community newsletter, which it distributed across the local area. The newsletter gave regular attention to BASF's "environmental achievements," as well as highlighting a variety of new community programs that the company had launched.[59] When Phil C. Greeson replaced Bill Moran as site manager in 1995, the BASF newsletter explained that the new manager would be making his home in Ascension Parish. Until Greeson's appointment, site managers had usually lived well away from the plant, but the company now felt that its most prominent executives should live in the community. "It's important to live close to where you work if you truly want to make a contribution to the community," noted Greeson. "Farmers learned a long time ago that they have to put something back into the soil if they hope to grow crops year after year. We, as corporate citizens, need to do the same thing. We want to put something back." Under Greeson's leadership, the plant also claimed that it wanted to hire increasing numbers of Ascension Parish residents.[60]

The change in BASF's approach was part of a broader shift. Since the late 1980s, rising pressure from citizens and environmental groups has pushed the industry to become increasingly vocal and sophisticated in fighting its corner. Under Dan Borne, who replaced Fred Loy as president in 1988, the LCA showed more responsiveness to public concerns about pollution. Borne, a polished public relations campaigner who had worked as Edwin Edwards' executive secretary in the 1970s, encouraged industry to improve its environmental performance. "The chemical industry got along because they employed people and paid taxes," he noted in 1992. "That was all that was expected. Today you have to do these things plus comply with environmental regulations."[61] Although making no mention of the dispute, the industry claimed that it made this realization "late in 1989," which was when the union's campaign was at its peak. At this time, the LCA set up the Louisiana Chemical Industry Alliance, which was made up of eight hundred members who came together to form a united front to address a variety of legislative and regulatory issues.[62]

Industry was also helped by a more favorable political climate. Buddy Roemer was detested by the chemical industry, and the LCA helped to resurrect Edwin Edwards to run against him in 1991. The election was complicated by the candidacy of David Duke, a notorious former Klansman who pushed Roemer into third place in the first gubernatorial primary. Anxious to cover his flank to the right, Roemer had switched to the Republican Party, a move that won him funding but lost him many of his environmental backers. His defeat was also blamed on his inability to connect personally with voters. Although Duke had run Edwards close in the first primary, once Roemer dropped out many of his supporters switched to Edwards in the runoff because they were afraid that a Duke victory might project a negative image of the state to the outside world. As a result, Edwards beat Duke by 61 percent to 39 percent.[63]

Under Edwards, the political climate became more favorable to industry. After Paul Templet refused to work for the new governor, Edwards scrapped many of his initiatives, including the environmental scorecard.

Taxes on imported waste were cut and penalties for noncompliance reduced. Toxic emissions, which had declined under Roemer, rose again soon after his defeat, although they were still lower than they had been prior to the lockout. By 1995, the EPA claimed that Louisiana had the worst water pollution record in the nation. Governor Mike Foster, who succeeded Edwards, continued many of his policies.[64]

In the decade after the lockout, African-American residents living near the chemical plants continued to bear the brunt of the pollution the plants produced. In 1993, for example, the Louisiana Advisory Committee to the U.S. Civil Rights Commission concluded that "black communities in the corridor between Baton Rouge and New Orleans are disproportionately impacted by the present State and local government systems for permitting and expansion of hazardous waste and chemical facilities."[65]

Throughout the decade, cancer rates in the communities along the chemical corridor remained high, ensuring that the "Cancer Alley" slogan remained in the news. The term was even used by Democratic candidate Bill Clinton during the 1992 presidential campaign to describe the pollution problems of Louisiana's chemical corridor.[66] Throughout the decade, industry and environmentalists continued to debate whether "Cancer Alley" was a reality. In November 1999, for example, investigative reporter Barbara Koppel's "Cancer Alley, Louisiana," published in the *Nation*, revisited many of the arguments that had been made a decade earlier. Residents living along the chemical corridor complained about high cancer rates while industry officials continued to stress lifestyle factors, particularly the prevalence of smoking and a fatty diet. Residents' groups hit back at industry claims, noting that a 1997 study by the U.S. Centers for Disease Control in Atlanta placed Louisiana in joint fifteenth place, among all fifty states, in the number of residents who smoke. They also argued that much of the population consumed large quantities of homegrown vegetables.[67]

As the twentieth century ended, it was still very difficult to conclusively prove the link between pollution and cancer rates, mainly because of the way that statistics are listed by the state's Tumor Registry. The registry divides the number of cancer cases from various parishes, some industrial and some rural, among ten regions. This makes it impossible to trace cancer rates close to the plants. Koppel also noted the extensive wealth of the oil and chemical industries and their large contributions to the state's universities and politicians, concluding that this provided the "biggest obstacle to proving the pollution-cancer link." The companies' biggest gifts, moreover, have been to the universities and medical centers that conduct cancer studies in conjunction with the State Health Department. Oil tycoon C. B. Pennington, for example, gave Louisiana State University $125 million in the 1980s to build a Bio-Medical Research Center.[68]

By the start of the twenty-first century, therefore, Local 4-620 and its environmental allies had made a concerted effort to tackle the state's environmental problems, but much remained to be done. Activists themselves recognized that the state still had many environmental problems, but they asserted that these would be much worse if they had not put pressure on industry. "I think the environmental community and the union folks have made a lot of progress," reflected Marylee Orr, "but because we produce so many chemicals, because we have the quote Cancer Alley or chemical corridor here and then over in Lake Charles, we still have way too much. So we still have a lot of work to do, but I think God help us if we all hadn't gotten involved, if all the communities and the labor union hadn't gotten involved, I'm certain it would be much much worse, no doubt."[69]

Conclusion

The union's survival in Geismar contradicts the images of defeat and decline often associated with organized labor in the 1980s and 1990s.[1] Since 1981, the decline of the American labor movement has been well documented. Private-sector union membership plummeted from 22.6 percent in 1981 to 14.1 percent in 1997, and the real wages of manufacturing workers suffered a corresponding fall.[2] Union decline began before the 1980s, but it was in the Reagan era that it began to seriously affect previously strong unions in industries such as autos, paper, steel, and construction.[3] In the 1980s, unions went down to a string of strike defeats, losing high-profile battles at Phelps Dodge, Hormel, International Paper, and Eastern Airlines. Labor unions, as one journalist noted in 1986, often seemed "virtually powerless to protect their jobs or their wages."[4] Many commentators indeed wrote organized labor off. Writing in 1991, labor lawyer Thomas Geoghegan described the union movement as "flat on its back." "'Organized labor,'" he lamented. "Say those words and your heart sinks. . . . It was a cause, back in the thirties. Now it is a dumb, stupid mastodon of a thing, crawling off to Bal Harbor to die."[5]

While it is universally recognized that since 1981 organized labor has suffered in a grim climate, it has also shown, as at Geismar, its ability to withstand powerful challenges. Labor's experience since 1981 cannot simply be expressed through defeat and decline; in the last twenty years, unions have also won some important victories. In 1989 and 1990, for example, the United Mineworkers revived militant traditions to head off an attempt by Pittston (Virginia) Coal Company to weaken their contract. Striking telephone workers in New York were similarly successful in resisting concessions demanded by NYNEX. In the 1990s, in particular, unions have continued to fight management concessions by adopting many of the same tactics that were used against BASF, especially forging bonds with community allies. In 1992, for example, a local union of the

United Steelworkers of America in Ravenswood, West Virginia, also secured a contract after enduring a lengthy lockout. The Ravenswood story, which a recent study has held up as a model for the "revival of American labor," relied on many of the same elements that had ensured the earlier success of the OCAW workers, particularly a broad-based corporate campaign, effective leadership, and rank-and-file militancy and participation. In 1997, the Teamsters' successful battle against UPS also illustrated the importance of these elements and proved to many observers that labor could not be written off quite so easily.[6]

On some occasions, labor bounced back from catastrophic defeats and showed an ability to turn management's hostile tactics to its advantage. In 1994, for example, Blue Circle Cement Company, a British multinational, hired permanent replacements to break a strike at its plant in Calera, Alabama. Once strikers had been replaced, the company then decertified the union. As they were gradually recalled to the plant, however, former union members began to sign up the replacement workers and eventually reorganized their union. The Calera story indicates that labor's experience in the 1980s and 1990s is not a simple, linear account of decline and defeat; there were also victories as well, and workers often exhibited surprising resilience.[7] Some recent studies have indeed revised the overwhelming emphasis on decline and have suggested a more balanced assessment of labor's position. Taylor Dark has urged readers to think "beyond the image of decline" and stressed labor's continuing political power, while other scholars have highlighted unions' recent responsiveness to the needs of women, African-Americans, and low-wage workers as cause for optimism.[8]

Some media coverage of the BASF dispute saw the OCAW's spirited corporate campaign as a sign that unions could not be written off as easily as many commentators had done. In May 1987, for example, the *Washington Post* viewed the OCAW's alliance with environmentalists as representative of a revitalized union movement that was willing to work with unusual allies in order to gain real leverage against companies. The *Post* noted that corporations that had taken on organized labor may "wish they had let sleeping dogs lie." A detailed article in the German newspaper *Die Weltwoche* also saw the OCAW's campaign as symbolic of the resurgence of American unions.[9]

The business press viewed the OCAW's campaign against BASF with some alarm. *Barron's* saw the effort as representative of unions' willingness to launch wide-ranging corporate campaigns against companies. The number of such campaigns, it noted, had risen from fewer than ten in 1983 to more than fifty in June 1987. The business publication felt that organized labor had found a dangerous new tactic, concluding that "unionism in this country once more is on the march." *Business International*, a weekly report circulated to managers of global companies, noted that the union's campaign highlighted a "new labor strategy" that would cause "corporate embarrassment" and "force targeted companies to expend large sums in public relations campaigns of their own." "A new era of labor militancy may be upon us," it warned, "differentiating itself from the 1930s and 1940s by its access to high technology communications and its international reach."[10]

Some of this coverage clearly exaggerated labor's recovery, and in subsequent years corporate campaigns have not always been successful. The BASF story itself showed that such campaigns were not a quick fix for the beleaguered U.S. union movement. The OCAW's struggle against BASF lasted for over five years and involved a massive commitment of time and resources. It enabled the union to make a principled stand against concessionary bargaining, but unions could not afford to devote the same amount of time and resources very often. "It took five years, and a lot of unions don't have the patience to build the coalitions we did," commented Richard Leonard.[11]

Some aspects of the BASF story were also unique. As a predominantly Catholic area with a French heritage, southern Louisiana is widely recognized as an unusual region within the South. Other aspects of the BASF workers' experience do offer suggestions for labor's recovery, however. The OCAW, a small union that only represented a minority of chemical workers, was successful in carrying on a sustained corporate campaign that was able to place real pressure on a multinational employer. BASF's plans to expand the Geismar site were crucial to the union's ability to use the regulatory process to exert leverage. The union's campaign highlighted corporate sensitivity to negative publicity, as well as showing that such efforts could succeed even without a consumer boycott.

Unity between the local and international union was also very important. The OCAW's leaders recognized the importance of the BASF fight and devoted a great deal of resources to the struggle. Unlike many conflicts of the 1980s, there was no criticism of the international union from OCAW members in Geismar; all were full of praise for the international union. In several disputes that labor lost, in contrast, including the strike at International Paper in 1987–88, local union members complained that the international leadership had not been sufficiently committed to the fight.[12] Richard Leonard remembered that the union's leadership gave him their full support throughout the corporate campaign. "Both Misbrener and Wages backed me one hundred percent," he recalled. "They gave me all the resources I needed, let me run loose, and didn't hamstring me with lawyers, because Bob was a lawyer, and as a lawyer he's got a lot of balls. Most labor lawyers are very tentative and afraid of getting sued, and Bob would say, 'Look man, rock and roll,' and we'd rock and roll."[13]

The union's success was also partly due to the effective alliance that it made with local African-American residents. Studies by environmental sociologists and others have brought to light the practice of locating polluting industries, including waste sites, in low-income, black communities. These communities were targeted partly because they had fewer abilities to engage in political protest, but in the early 1980s their residents confounded expectations and started to exhibit a growing militancy against industrial polluters. The experience of residents living in the Geismar area therefore fits into this broader pattern of emerging black environmental protest. Across the South, African-American communities pioneered a new "environmental equity" movement as they came to feel, as in Geismar, that their communities were being subjected to an excessive amount of industrial pollution. In Houston, for example, African-American residents organized against the placement of waste disposal facilities in their neighborhoods, while in Triana, Alabama, local African-Americans waged a four-year court battle against an insecticide manufacturer in their town after tests by the U.S. Centers for Disease Control found that residents had very high levels of insecticides in their blood.[14]

Like the residents of Geismar, who were led by a former civil rights activist, many of these communities also drew on their civil rights experience to wage new battles for a cleaner environment. In Warren County, North Carolina, black church leaders led nonviolent civil disobedience protests to prevent trucks from entering a landfill located in a predominantly African-American community. The ministers involved in the Warren County protests included Benjamin Chavis, who later became the executive director of the NAACP. Minority communities outside the South also fought back; in the mid-1980s, African-American and Hispanic residents in South Central Los Angeles, for example, joined forces to stop a large incinerator from being built in their community. In 1991, more than 650 grassroots environmental activists from across the country, including representatives from Louisiana, converged on Washington for the first People of Color Environmental Leadership Summit, a gathering that pushed established environmental groups to begin addressing charges that they had failed to reach out effectively to minority communities.[15]

OCAW leaders viewed the BASF dispute as a model of success that they hoped would turn the fortunes of the union around.[16] These hopes have, however, only been partially realized. The lockout did not stop the OCAW's continued membership decline, although the pace of this decline was less rapid after the dispute than it had been in the late 1970s and early 1980s. Between 1979 and 1984, for example, the OCAW's membership plummeted from 180,017 to 118,301. This precipitous drop reflected the impact of both concessionary bargaining and plant layoffs and shutdowns. By 1994, in contrast, the number of members had declined less slowly, to around 90,000.[17] OCAW leaders claimed that the BASF dispute helped to slow the pace of membership decline by showing industry that the union was capable of resisting concessionary bargaining. Robert Wages, elected president in 1991, asserted in 2000 that "Our success with BASF gave pause to the other multinationals that we dealt with, in the oil industry and the chemical industry, that we were capable of a pretty destructive fight."[18]

It was natural for OCAW leaders to stress the importance of the BASF dispute, yet even this fight could not stop the reality that the union was continuing to lose members throughout the 1990s, albeit less quickly than before. Plant closings and downsizing were cited by union leaders as the major causes of these losses. "We were hurt by virtue of deindustrialization," noted OCAW leader Tony Mazzochi. This was certainly a cause of the union's decline, but the OCAW was also slow to organize new members to replace those that it was losing, only setting up an organizing department in 1994. "We didn't really tackle the core reasons [for decline], the core problem of organizing until the early nineties," admitted Richard Leonard. "Actually 'ninety-three, 'ninety-four was when we really started to address it as an organization, thoughtfully address it. . . . Other unions in the industrial sector probably got a ten-year jump on us."[19] By the end of the 1990s, the OCAW's membership had dipped to around 82,000. Not for the first time, the union's leaders began to look for a merger partner, initiating talks with the United Paperworkers International Union, a much larger union, with over 200,000 members. In 1999, these talks led to the creation of the new Paper, Allied-Industrial, Chemical, and Energy Workers' International Union (PACE).[20]

Like the local union, after the lockout the national OCAW also made a conscious effort to strengthen the alliances it had made during the dispute. The union has continued to oppose the increasing use of contractors in the petrochemical industry, asserting that these workers threaten safety standards. In June 1990, the union released *Out of Control*, a video documentary produced in conjunction with Ralph Nader. Consisting of interviews with plant workers and industry experts, the film claimed that the increasing use of contract labor was linked to chemical catastrophes such as the 1989 explosion at a Phillip's Refinery in Houston that killed 23 workers and injured another 272. Robert Wages, who supervised the distribution of *Out of Control* across the United States, described it as part of an effort "to build bridges between labor and community strong enough to bear the weight of a sweeping program of reform at local, state, and national levels. . . . Basically, we're capitalizing on our experiences in Geismar and elsewhere and institutionalizing them on a national level."[21]

In the decade since the ending of the BASF dispute, the OCAW has also continued to emphasize environmental issues in its corporate campaigns. In 1994, for example, the union asserted that an Occidental Chemical Corporation plant in Belle, West Virginia, was violating environmental regulations. OCAW leaders believed that Occidental did not want to deal with the union and that this had influenced its decision to shut the plant. Following an investigation into the union's complaints, the state Division of Environmental Protection cited the company for forty-nine hazardous waste violations. The decision did not stop the company from closing the plant, but it did ensure negative publicity for Occidental as well as giving some satisfaction to the laid-off OCAW members.[22]

The environment also became a central issue in a lengthy dispute against Crown Central Petroleum in Pasadena, Texas. After the company locked out OCAW members in 1996, the union responded by working with local residents who were concerned about pollution from the plant, filing a lawsuit against Crown for violations of the Clean Air Act. As was the case in the BASF dispute, both the residents and the locked-out workers insisted that the plant's pollution record became worse when the refinery was operated by temporary contract workers. In August 1998, the union's position was partially upheld when Crown received a $1,055,425 fine from the Texas Natural Resource Conservation Commission, the largest air pollution penalty in the state's history. More than two years later, however, the Crown workers were still locked out of their jobs.[23]

In recent years, some scholars have noted examples of cooperation between labor and environmentalists, finding that the amount of conflict between the two groups has been exaggerated. Scott Dewey has shown that organized labor demonstrated "relatively strong support for many environmental initiatives prior to 1970," particularly at the leadership level. Long-serving UAW leader Walter Reuther, for example, gave firm backing to a wide range of environmental initiatives. Under Reuther, the UAW established a Department of Conservation and Resource Development in 1967 and supported stricter legislation to provide clean air and water.[24] By the 1970s, in particular, some labor leaders realized that conflict between unions and environmentalists generally served the interests of industry. The two groups could join together to secure a strong economy and a healthy environment, they argued, especially as improved environmental protection laws could create more jobs than they destroyed. Political scientist Benny Temkin has pointed out that during the energy crisis of the 1970s, several unions made coalitions with environmentalists, the two groups finding common ground in their mutual opposition to large corporations.[25] Robert Gordon has explored the close cooperation between the OCAW and environmentalists during the 1973 strike against Shell Oil Company, while other environmental historians have called for more "bridge-building between the fields of environmental and social and labor history."[26]

While the BASF lockout has not previously been written about in a book-length study, several environmental scholars and activists have noted its importance as a prominent example of successful cooperation between organized labor and the environmental movement. In *Forcing the Spring: The Transformation of the American Environmental Movement*, Robert Gottlieb notes that the union's "redefinition of the work/environment relationship" suggests "new strategies [that] can form the basis for new kinds of social movements."[27] Robert Gordon asserts that the lockout led to a "revitalization" of the alliance that organized labor and environmentalists had forged during the Shell strike, while Jim Schwab sees the growth in environmental consciousness among Local 4-620's members as part of the broader emergence of blue-collar and minority environmentalism. Lois Gibbs, one of the lead activists in the efforts to clean up the Love Canal dump in the late 1970s, has also written of the lockout as an inspiring tale that confirms that workers and environmentalists could find common ground.[28]

The BASF lockout certainly showed that even in the 1980s, a decade when both movements suffered in a conservative political climate, organized labor and environmentalists could work together effectively. At the same time, however, it showed what hinders the two movements from cooperating more often. Workers' fear that environmentalists threaten their jobs has been a particularly difficult barrier to overcome. The BASF workers have continued to be supportive of environmentalists because they have been through the lockout, an experience that permanently transformed their environmental views. Workers who had not experienced such a dispute, however, were generally much more supportive of industry, just as the BASF workers themselves had been before the lockout. Even other OCAW locals had doubts about working with environmentalists. The "tricky problem," noted Richard Miller, was that "chemical industry workers perceive the agenda of environmentalists as a threat to their jobs unless they are at war with management."[29]

Since the BASF dispute, some unions have shown very little interest in working with environmentalists, either within Louisiana or at a national level. The United Mineworkers, in particular, has continued to place more priority on job security for its members than on environmental protection.[30] The lockout also showed that both environmentalists and unions risked alienating some of their own core support if they worked together. Darryl Malek-Wiley's alliance with OCAW, for example, cost him support among some Sierra Club members who saw unions as "the enemy." The OCAW's alliance with environmentalists also reduced the amount of backing they received from other unions within both the United States and West Germany.[31]

Since the mid-1980s, however, there have been other examples of cooperation between organized labor and the environmental movement. Within Louisiana, the dispute demonstrated to the leaders of LEAN the value of working with organized labor, and the statewide environmental organization now has three unions represented on its board. Outside the Pelican State, groups such as the Labor-Community Strategy Center in Los Angeles have also worked to redirect both movements toward a common understanding. By the mid-1990s, there were also signs that the national leaders of the AFL-CIO were starting to recognize the value of working with environmentalists. Labor and environmental organizations joined together to oppose the 1993 North American Free Trade Agreement because they claimed that it lacked strong protections for labor and the environment. In 1995, newly elected AFL-CIO president John Sweeney appointed Jane Perkins, a former president of Friends of the Earth, to be the labor federation's first liaison to the environmental movement. Sweeney also created an environmental-policy committee on the AFL-CIO executive council and gave his support to Just Transition, a program that proposed a government-established fund to compensate workers who were laid off when environmentally unfriendly products were phased out. Originally known as a Superfund for Workers, Just Transition was formulated in the 1980s by the OCAW's Tony Mazzochi and organizers with the National Toxics Campaign. Mazzochi, a World War II veteran, modeled his program on the GI bill, which had provided living allowances and tuition payments to veterans heading for college. "I kept insisting we ought to use what we learned after World War II for transition, because that was a great transitional program which society greatly benefited from," he recalled.[32]

The Just Transition program provides a way of mobilizing workers behind environmental initiatives by directly addressing their fear that environmentalists would simply shut industry down without giving them any compensation. In the 1990s, the OCAW also ran a series of workshops in which it brought workers and environmentalists together to try and confront their mutual prejudices. Both sides agreed that the workshops helped to break down misconceptions that the two groups often had about each other, highlighting that education could change their views. "We don't want them to go out and eliminate our jobs," noted one refinery worker, "but I understand where they're coming from. We have the same feelings."[33]

A decade after the settlement of the dispute, many of the leading local figures from the lockout no longer lived in or stayed connected to Louisiana. Les Story stayed with BASF until the early 1990s, when he left to work for another chemical manufacturer. In the fall of 2000, Story insisted that he had enjoyed his time in Louisiana, and he stood by the decision to subcontract the maintenance work. He felt that the union had made a "mistake" by "allowing the productivity of their mechanics to slide," asserting that the introduction of contract maintenance had helped the site to thrive since the lockout. By 2000, both Richard Donaldson and Bill Jenkins, Story's management colleagues, had retired from their careers in the chemical industry. Donaldson still lived in Baton Rouge, but Jenkins had retired to the mountains of East Tennessee, where he had been raised.[34]

Richard Leonard continued to work for the OCAW after the lockout, running corporate campaigns and coordinating organizing efforts. Richard Miller stayed in Louisiana for two years after the end of the dispute and helped establish the LLNP. With a partner in Massachusetts, however, Miller looked for a chance to head north, although he had grown fond of Louisiana and the friends he had made there. Toward the end of 1991, he left the state to work as a Washington, D.C.–based consultant for the OCAW and later PACE. For the rest of the 1990s, Miller tackled his new job with the same dedication that he had shown during the lockout.[35]

In 1992, Esnard Gremillion was awarded the Samuel Gompers Union Leadership Award by the AFL-CIO for his work during the lockout. He traveled to Washington to receive the prize, which was given to five "enterprising trade unionists" each year. After heading the union's campaign for nearly six years, Gremillion settled down to a life of retirement for the rest of the 1990s. At the end of 2000, he lived a quiet life in the suburbs of Baton Rouge, surrounded by his large extended family.[36]

One of the leading figures from the lockout became a casualty of cancer a few years after the dispute had ended. In November 1993, John Daigle, who had served as president of Local 4-620 since 1977, became ill and was diagnosed with pancreatic cancer. The cancer spread rapidly, and Daigle died on February 23, 1994, at the age of fifty-nine. His coffin was carried by Local 4-620 members who had been through the lockout with him, including Bobby Schneider, Tommy Landaiche, and Esnard Gremillion. Some union members linked Daigle's death to chemical exposure at the plant, where he had worked since 1958. "As a long time chemical plant worker, I know years have been trimmed off of my life from exposure to chemicals," Daigle himself had commented in 1989. "I have seen plants vent chemicals at night after the DEQ inspectors have gone home to their families." Daigle's loss was felt keenly by Local 4-620's members, who erected a small memorial in the union hall in his honor. "Friend and Brother Forever" it read. The 1994 Louisiana AFL-CIO convention was also dedicated to Daigle. He was succeeded by Duke King, who served as local union president for the rest of the 1990s.[37]

In 1999, Local 4-620 moved to a new union hall in Gonzales that was funded through union members' contributions. The smart new building was designed to include office space for the LLNP. The move from Baton Rouge to Ascension Parish highlighted the union's increased awareness of the needs of local residents, as well as making it easier for workers to visit the building after their shifts. Within the hall, the LLNP's offices were next to those of the local union officers, underscoring the continued partnership of the two groups. Union members also took care to construct an outdoor cooking area, allowing them to expand their jambalaya cookouts

24. John Daigle, pictured in 1989. (Courtesy PACE International Union)

that were so popular with members. A large meeting area was decked out with memorabilia from the lockout, evidence of the pride that most workers felt about the their action. Many saw the hall as a monument to their struggle in the lockout, particularly to those who had died during that time.

More than a decade after the settlement of the BASF dispute, the union's billboards on the interstate were a distant memory, partly because BASF had purchased the land where the signs had been placed. Keen to prevent the union from having such a public platform for its allegations, the company had reportedly paid the union member who owned the land well above the market price for it. Trucks coming to the chemical complex in Geismar were still, however, avoiding highways 73 and 74, as they had agreed to do in 1987, and a sign on the Geismar exit reminded them that "No hazardous materials" were allowed through the small riverside community. By 2000, Amos Favorite still led the environmental organization that had secured this change. He was now seventy-seven years old and not as physically active as he had been, but he felt pride in the achievements of the APRATP, particularly the bond it had forged with the OCAW. "We know what's going on, and we know how to fight it," he reflected. "We know where to fight it at, and all that kind of stuff, and then we have our allies that work in them plants. . . . It improved a lot, but no I wouldn't say that we got it a hundred percent, but it's improved a lot. . . . but we brought it down a long ways, me and Willie Fontenot and a bunch of guys that was determined to stop it, because man, the way them people was going, we wouldn't have had no more civilization here. People would have all died out from the way the chemical releases was going on around here."[38]

Notes

Introduction

1. For an insight into this popular notion of conflict, and the way that it is now being increasingly revised, see Kazis and Grossman, *Fear at Work*, x; Gordon, "Shell No!" 460–1; Dewey, "Working for the Environment," 45; Montrie, "Expedient Environmentalism," 75–6.

2. Schwab, *Deeper Shades of Green*, 234.

3. For historians' views that Reagan's treatment of the PATCO strikers encouraged companies to take a hard line with their unions, see, for example, Zieger, *American Workers*, 198; Berman, *America's Right Turn*, 98.

4. Zieger, *American Workers*, 193–5, 198–200.

5. Dark, *The Unions and the Democrats*, 15.

6. Winston Williams, "Business Brings Back the Lockout," *New York Times*, October 5, 1986, clipping in "Media/Clips/Statistics" folder, box 15, OCAW International Union Papers, PACE (Paper, Allied-Industrial, Chemical, and Energy Workers' Union), Nashville (hereinafter cited as IUP). It is difficult to statistically document the increasing number of lockouts because the federal government kept no accurate lockout statistics, grouping them with strikes as labor disputes. Both management, labor, and economists, however, universally agreed that use of the lockout weapon was on the increase, and the trend was widely commented on in the national media throughout the 1980s. Much of the press saw the increasing prevalence of lockouts as a return to the labor conflict of the Great Depression era. In 1989, for example, the *Washington Post* also commented on the increasing frequency of lockouts in an article detailing the end of the BASF dispute. See Frank Swoboda, "5½-Year Labor Lockout Ends," *Washington Post*, December 16, 1989, clipping in "Labor/End Lockout-Settlement" folder, box 6, IUP. For other examples, see Robert Walters, "Some Businesses Using Strong-arm Tactics," *Lumberton (N.C.) Robesonian*, September 17, 1987, clipping in "1987" folder, untitled box, OCAW Local 4-620 Papers, PACE Local 4-620, Gonzales, La. (hereinafter cited as LUP); "Anti-union Lock-outs Increasing," *Augusta (Ga.) Herald*, August 26, 1987, clipping in "1987" folder, untitled box, LUP; Robert Walters,

"Business Flexes Its Muscle: Use of Lockouts on the Rise," *Norristown (Pa.) Times Herald,* September 4, 1987, "1987" folder, untitled box, LUP.

7. Both labor and management widely referred to the BASF dispute as the longest labor lockout in American history. See, for example: "Workers Vote on Plan to End BASF Dispute," *New Orleans Times-Picayune,* December 19, 1989, p. D1; Joseph M. Misbrener et el. to All OCAW Local Unions, December 7, 1990, "Post Lockout," box 11, IUP; John E. Mullane to Robert G. Thoma, July 9, 1987, Story Papers.

8. "'Coordinated Corporate Campaigns' and the Local Labor Dispute at BASF's Geismar Works Plant: A Backgrounder," March 6, 1986, "Corporate Campaign Effectiveness" folder, box 2, IUP.

9. Jonathan Tasini, "Labor Unions Fine-Tune Battle," *New York Newsday,* December 31, 1989, clipping in "BASF General News Items" folder, box 3, LUP.

10. Richard Leonard to Don Southard, May 27, 1988, "1988 Correspondence" folder, box 2, IUP.

11. "Locked-Out Workers Go Public to Pressure BASF," *Multinational Monitor,* March 15, 1986, pp. 10–1.

Chapter 1. The Trade-Off

1. "Prosperity in Paradise?: Louisiana's Chemical Legacy," *Baton Rouge Morning Advocate,* April 25, 1985, supplement, p. 3, in "Locations/Louisiana" folder, box 5, IUP.

2. Susan Reed, "His Family Ravaged by Cancer, an Angry Louisiana Man Wages War on the Very Air That He Breathes," *People Weekly,* March 25, 1991, pp. 42, 43, 44, 48; transcript of *Locked Out!* video, October 7, 1988, p. 3, "BASF v. OCAW Correspondence No. 3" folder, box 12, IUP.

3. Borne interview.

4. "Partners in Louisiana: Working for Louisiana," LCA Publication, pp. 39–40, in "Chemical Industry" file, Ascension Parish Library (quotation on p. 40).

5. Bartley, *The New South,* 401–2; Schulman, *From Cotton Belt to Sunbelt,* 206–18; Borne interview; "Gov. Edwards and Industrialists," *New Orleans Times-Picayune,* January 28, 1973, sec. 1, p. 16; "No More Tax Increases Seen," *New Orleans Times-Picayune,* January 26, 1973, sec. 1, p. 1; "Union Carbide Will Expand," *New Orleans Times-Picayune,* August 16, 1974, sec. 1, p. 5.

6. "Gov. Edwards and Industrialists," *New Orleans Times-Picayune,* January 28, 1973, sec. 1, p. 16; "Chemical Industry a Bulwark," *New Orleans Times Picayune,* January 12, 1975, sec. 1, p. 20.

7. "Prosperity in Paradise?: Louisiana's Chemical Legacy," *Baton Rouge Morning Advocate,* April 25, 1985, supplement, p. 3, in "Locations/Louisiana" folder, box 5, IUP; "DuPont Buys LaPlace Site: $120 Million Development Seen," *Gonzales Weekly,* January 25, 1957, p. 1; *Gonzales Weekly,* May 16, 1958, p. 10.

8. Gilin interview; "Wyandotte Plant at Geismar Gets into Operation," *Gonzales Weekly,* June 20, 1958, p. 1; "Wyandotte Dedicates New Chemical Plant,"

Gonzales Weekly, June 27, 1958, p. 1; "Wyandotte Chemical Announces Major Construction at Geismar, *Gonzales Weekly,* March 8, 1968, p. 1.

9. Donaldson interview.

10. Transcript of *Locked Out!* video, October 7, 1988, p. 1, "BASF v. OCAW Correspondence No. 3" folder, box 12, IUP.

11. Gremillion interview; Smith interview.

12. Davidson, *Challenging the Giants,* 243–4; Wages interview.

13. Rousselle interview; Richard Leonard to Michael Silverstein, December 1, 1986, "1986 Correspondence" folder, box 2, IUP; Gremillion interview.

14. "Du Pont Awaits Another Organizing Election," *Chemical Week,* December 2, 1981, p. 14; "Du Pont Employees Turn Down the Steelworkers," *Chemical Week,* December 23, 1981, p. 14.

15. "Du Pont Is the Target," *Chemical Week,* February 23, 1972, p. 17.

16. Schneider interview.

17. Rousselle interview. For many examples of grievances, see "1976 Grievances" folder, untitled box, LUP; "1978 Grievances" folder, untitled box, LUP.

18. Hawkins interview.

19. Braud interview; Fink interview; Schneider interview.

20. "Geismar Strike Continues," *WyChem News,* November 1970, p. 3, copy in "BASF Info Pre-1975" folder, box 1, IUP; "New Contract Signed at Geismar Works," *BASF Wyandotte News,* March 1971, p. 1, copy in "BASF Info Pre-1975," folder, box 1, IUP; Rousselle interview; Jenkins interview.

21. Rousselle interview.

22. Rousselle interview; Jenkins interview.

23. "Corporate Campaign Replaces Strike as Primary Union Tool," *Business International,* September 22, 1986, clipping in "Media/Clips/Stats" folder, box 15, IUP; John Kirkman, letter to author, July 19, 2001, copy in author's possession; King interview.

24. Pederson, *International Directory,* 47.

25. Ibid., 47–8.

26. Ibid., 47–8; Adam Lebor, "War Crimes: Have Germany and America Sold Out on the Holocaust Survivors?" *Times* (London, magazine supplement), October 28, 2000, pp. 24, 25, 27, 28, 30.

27. BASF Corporation Factsheet, n.d., Story Papers.

28. BASF Corporation Factsheet, n.d., Story Papers; Donaldson interview; Jenkins interview.

29. "Polyol Plant to Be Built At Geismar," *BASF Wyandotte News,* February 1971, p. 1, copy in "BASF Info Pre-1975" folder, box 1, IUP; "BASF Corporation's Aniline Plant in Geismar, Louisiana," "BASF General" folder, Louisiana Environmental Action Network Papers (hereinafter cited as LEAN Papers); "Fact Sheet—BASF Wyandotte Geismar Works," December 3, 1985, Story Papers; Story interview; John Hall, "Firm Will Build Plant in Geismar," *New Orleans Times-Picayune,* March 30, 1992, p. D2.

30. Arnold interview.

31. "Chemical Industry a Bulwark," *New Orleans Times-Picayune*, January 12, 1975, sec. 1, p. 20.

32. "BASF Corporation Chemicals Division: Plant Seniority List," February 18, 1987, Story Papers.

33. Hawkins interview; Schneider interview; John Daigle, quoted in "Swedish TV film" videotape, copy in LUP.

34. Transcript of *Locked Out!* video, October 7, 1988, p. 16, "BASF v. OCAW Correspondence No. 3" folder, box 12, IUP.

35. Hawkins interview.

36. Fink interview; Gremillion interview.

37. Carson, *Silent Spring;* Graham, *Since Silent Spring*, 13–7; Gordon, "Shell No!" 466.

38. Goldfarb, "Environmental Legislation," 547.

39. Merchant, introductory section in *Major Problems*, 523; Goldfarb, "Environmental Legislation," 548, 549.

40. Merchant, introductory section in *Major Problems*, 523. As studies by environmental sociologists have shown, the persistence of public concern about the environment through the Reagan years highlighted that environmental quality was an "enduring" concern for many Americans. These studies have also shown, however, that proenvironment responses to opinion polls did not translate directly into proenvironment votes. In the 1984 presidential election, many voters were concerned about cuts in environmental programs, but Reagan's association with economic recovery was a decisive factor in his reelection. Dunlap, "Public Opinion," 7–8, 35–6 (quotation on p. 7); R. Anthony, "Trends in Public Opinion," 14–5, 19–20. For a detailed insight into the Reagan administration's environmental policies, see Vig and Kraft, eds., *Environmental Policy.*

41. Arnold interview; Schneider interview.

42. Davidson, *Peril on the Job*, 188.

43. Gordon, "Shell No!" 466.

44. Kazis and Grossman, *Fear at Work*, 3.

45. Ibid., x–xi (quotation on p. x).

46. Mazzochi interview.

47. "La. Industrial Air Penalties 'Drop in Bucket,'" *Baton Rouge Morning Advocate*, April 25, 1985, supplement, p. 11, in "Locations/Louisiana" folder, box 5, IUP; "Sportsmen Lose Paradise, Piece by Piece," *Baton Rouge Morning Advocate*, April 25, 1985, supplement, p. 11, in "Locations/Louisiana" folder, box 5, IUP; Colten, "Texas v. the Petrochemical Industry," 148.

48. Borne interview; Donaldson interview.

49. Roemer interview; "Edwards: 'Yes, We Made Trade-offs,'" *Shreveport Times*, May 6, 1990, clipping in "Louisiana's Environment: Heritage Held Hostage" folder, LEAN Papers (quotation on p. 1).

50. Badger, "When I Took the Oath," 2, 4, 8.

51. Ibid. (quotation on p. 4).

52. Bullard, *Dumping in Dixie*, 1–2; C. Anthony, "Why African-Americans," 541–2. For an overview of studies by environmental sociologists that have explored black environmental attitudes, see D. Taylor, "Blacks and the Environment."

53. Hasten interview.

54. Minchin, "Federal Policy," 148–9; Brattain, *The Politics of Whiteness*, 35–7.

55. F. R. Donaldson and C. V. Kirkland to Charles R. Burl, November 13, 1972, "Justice Department Suit 1972" folder, untitled box, LUP; Smith interview; Landaiche interview.

56. David L. Norman to John D. Blodger, April 22, 1972, "Justice Department Suit 1972" folder, untitled box, LUP.

57. Consent decree for *United States of America v. BASF Wyandotte Corporation*, July 4, 1972, "Justice Department Suit 1972" folder, untitled box, LUP; Donaldson interview.

58. "BASF General News Items" folder, box 3, LUP; Smith interview.

59. "Up from the Chemical Plantation," p. 62, "Media/Interview/Leonard/Benman" folder, box 9, IUP.

60. Harvey interview.

61. "OCAW Strategy for BASF Campaign," July 13, 1984, box 16, IUP; Fact Sheet—BASF Wyandotte Geismar Works, December 3, 1985, Story Papers; Donaldson interview.

Chapter 2. Negotiations

1. Schneider interview.

2. Donaldson interview.

3. Getman, *The Betrayal of Local 14*, 10–1, 16–7; Brody, "The Breakdown of Labor's Social Contract"; Nissen, "A Post–World War II 'Social Accord'?" 173–4.

4. Zieger, *American Workers*, 198.

5. Employers' right to hire permanent replacements originated in the *MacKay Radio v. National Labor Relations Board* Supreme Court decision of 1938. The decision, while ordering that several workers be reinstated in this particular instance, volunteered language in its decision that it is not an unfair labor practice to permanently replace striking workers. Prior to the 1980s, however, the use of permanent replacements was very rare and was mainly limited to smaller companies not susceptible to widespread public pressure. The position of unions was slightly strengthened in the late 1960s when the NLRB ruled in the *Laidlaw* case that economic strikers who are permanently replaced are, after the end of the dispute, entitled to reinstatement when openings occur among the new workforce. See Philip Mattera to Corporate Campaign, November 14, 1988, "Replacement Workers" file, United Paperworkers' International Union Papers; Peter T. Kilborn, "Replacement Workers: Management's Big Gun," in Boris and Lichtenstein, eds., *Major Problems*, 598–600.

6. Zieger, *American Workers*, 200.

7. Story interview; Donaldson interview. All the BASF managers I interviewed stressed that the U.S. arm of the company set labor relations policies. This was also confirmed by BASF executives in Europe. See Nicola Palmieri-Egger, letter to author, January 12, 2001, copy in author's possession.

8. Jenkins interview; "E. Stenzel Becomes President, Succeeding Dieter H. Ambros," *BASF Wyandotte News*, April 1979, pp. 1, 3, copy in "75–79" folder, box 1, IUP. Mr. Stenzel did not respond to requests to be interviewed for this project.

9. Jenkins interview; Donaldson interview.

10. Henry Kramer, letters to author, December 11, 2000, and February 7, 2001, copies in author's possession.

11. "The 'Jelly Bean' Theory: Strategies in Dealing with Unions," April 26, 1983, Story Papers (quotation on p. 25). Les Story was also quite open about BASF's preference for a nonunion operation. "We never believed that any of our employees needed to be represented by somebody, by a third party," he explained; ". . . The entire company felt that way." Story interview.

12. E. F. Schuknecht to L. J. Story, June 28, 1983, Story Papers.

13. Richard Leonard to Chris Bedford, March 7, 1988, "1988 Correspondence" folder, box 2, IUP.

14. Wages interview; Jenkins interview; Henry Kramer, letter to author, February 6, 2001, copy in author's possession.

15. "The 'Jelly Bean' Theory: Strategies in Dealing with Unions," April 26, 1983, p. 26, Story Papers.

16. Jenkins interview; 274 NLRB No. 147, pp. 978–89 (quotation on p. 988).

17. "Initial Proposals from BASF Wyandotte Corporation to OCAW Local 4-620," May 14, 1984, LUP.

18. "OCAW Dispute with BASF Continues in Louisiana," *Daily Labor Report*, November 9, 1987, copy in "Media/Clips/Statistics" folder, box 15, IUP; Jenkins interview.

19. Boyer et al., *The Enduring Vision*, 717; Norton et al., *A People and a Nation*, 647.

20. Story interview; Wage Rate Data, May 1, 1984, Story Papers.

21. Rousselle interview; Story interview.

22. "'Coordinated Corporate Campaigns' and the Local Labor Dispute at BASF's Geismar Works Plant: A Backgrounder," March 6, 1986, pp. 9–10, "Corporate Campaign Effectiveness" folder, box 2, IUP; Jenkins interview.

23. Arnold interview; Braud interview; Rebecca Wiltz, "What Is the Women's Support Group?" n.d., untitled folder, untitled box, LUP.

24. "OCAW Workers to Go Abroad to Pursue Dispute against German Chemical Firm," *DLR*, January 17, 1985, p. A-8, copy in "Labor-Negotiations" folder, box 6, IUP.

25. "Negotiations Meeting between the Representatives of BASF Wyandotte Corporation, Geismar, Louisiana, and Oil, Chemical and Atomic Workers, Local 4-620," June 1, 1984, p. 20, LUP.

26. "Negotiations Meeting between the Representatives of BASF Wyandotte Corporation, Geismar, Louisiana, and Oil, Chemical and Atomic Workers, Local 4-620," June 13, 1984, p. 25, LUP.

27. Rousselle interview; Jenkins interview.

28. Story interview.

29. Gremillion interview; Arnold interview. For similar opinions, see also Smith interview; Landaiche interview.

30. Roemer interview; Guste interview.

31. Rousselle interview; Henry Kramer, letter to author, December 11, 2000, copy in author's possession.

32. Jenkins interview; Story interview.

33. "The 'Jelly Bean' Theory: Strategies in Dealing with Unions," April 26, 1983, pp. 5–19, Story Papers (quotations on pp. 6, 19).

34. "Council on Union-Free Environment: An Organization for Positive and Progressive Employee Relations," May 16, 1982, Story Papers; "Positive Planning for Union-Free Operations," May 27, 1983, Story Papers; Fred K. Foulkes, "How Top Nonunion Companies Manage Employees," *Harvard Business Review*, September-October 1981, pp. 90–6, copy in Story Papers.

35. L. J. Story to E. F. Gremillion, February 26, 1985, "Labor-Negotiations" folder, box 6, IUP.

36. Transcript of *Locked Out!* video, October 7, 1988, p. 4, "BASF v. OCAW Correspondence No. 3" folder, box 12, IUP; Jenkins interview.

37. "Emergency Work Pay Policy," March 5, 1984, "Labor-Negotiations" folder, box 6, IUP.

38. "A Plan to Rescue the Louisiana Chemical Industry," *Chemical Week*, January 11, 1984, p. 10.

39. "Managing during the Recession," *Chemical Week*, January 20, 1982, p. 5.

40. "A Chemical Union Gets a New Chief at a Troubled Time," *Chemical Week*, January 4, 1984, pp. 10–1 (quotations on p. 11).

41. See, for example, "A Troubled OCAW Looks to a Three-Way Merger," *Chemical Week*, May 23, 1984, p. 4.

42. "Industry's Tough Line on Cutting Labor Costs," *Chemical Week*, February 3, 1982, p. 49; "The Teamsters Settle for Less," *Chemical Week*, January 27, 1982, pp. 13–4 (quotation on p. 13).

43. Story interview.

44. *How Louisiana Passed Right to Work* (Baton Rouge: Louisiana Association of Business and Industry, 1977), copy in author's possession (quotations on pp. 21, 33); Steimel interview.

45. *How Louisiana Passed Right to Work*, i; Steimel interview.

46. Juravich and Bronfenbrenner, *Ravenswood*, 9.

47. Joseph M. Misbrener to Brothers and Sisters, January 6, 1987, "Correspondence 1987" folder, box 2, IUP.

48. "Negotiations Meeting between the Representatives of BASF Wyandotte

Corporation, Geismar, Louisiana, and Oil, Chemical and Atomic Workers, Local 4-620," June 13, 1984, p. 28, LUP.

49. Story interview.

50. "Negotiations Meeting between the Representatives of BASF Wyandotte Corporation, Geismar, Louisiana, and Oil, Chemical and Atomic Workers, Local 4-620," June 15, 1984, p. 20, LUP.

51. Ibid.

Chapter 3. They've Got Us Cornered

1. Miller interview.

2. Rousselle interview; L. J. Story to E. F. Gremillion, October 24, 1984, "Labor-Negotiations" folder, box 6, IUP.

3. Fink interview; Smith interview; Hawkins interview; Harvey interview.

4. Arnold interview.

5. Rebecca Wiltz to Representative Cathy Long, July 5, 1985, "Women's Support Group" folder, box 4, LUP.

6. Rousselle interview.

7. Hawkins interview; Landaiche interview.

8. Wiltz interview.

9. Transcript of address by Connie Kearns to the Louisiana AFL-CIO Convention, March 18–22, 1985, untitled folder, "Richardson" box, LUP.

10. See, for example, Rebecca Wiltz to Representative Cathy Long, July 5, 1985, "Women's Support Group" folder, box 4, LUP.

11. "Women's Support Group," untitled box, LUP; Rebecca Wiltz, "What Is the Women's Support Group?" n.d., untitled box, LUP.

12. Juravich and Bronfenbrenner, *Ravenswood*, 46.

13. Wiltz interview.

14. Ibid.

15. Nordstrom interview; Wiltz interview; Rousselle interview; Gremillion interview.

16. Wiltz interview; Nordstrom interview.

17. Harvey interview; Rebecca Wiltz, "What Is the Women's Support Group?" n.d., untitled box, LUP.

18. Braud interview; Smith interview.

19. EEOC Charge of Arthur J. Walker, May 11, 1984, EEOC Charges of Roosevelt Leonard, Floyd McGalliard III, and Perry I. Taylor, May 11, 1984, EEOC Charge of Frank L. Benjamin, April 13, 1984, "EEOC Charges" folder, box 4, LUP.

20. Smith interview; Favorite interview; Miller interview; "Death Takes Three Members of Local 4-620," n.d., "Corporate Campaign" folder, box 4, LUP. For the broader racial climate of the late 1950s (when Local 4-620 was chartered), and how it affected organized labor in the South, see Draper, *Conflict of Interests*, esp. 17–40.

21. "Assessment of What 'Lockout' Was Costing Plant," July 17, 1984, Story Papers.

22. Gilin interview.

23. Getman, *The Betrayal of Local 14*, 47–54; Zieger, *American Workers*, 198–9.

24. Richard Donaldson to author, January 20, 2001, copy in author's possession.

25. Guidry interview.

26. "State Jobless Rate Still High," *New Orleans Times-Picayune*, March 25, 1984, sec. 1, p. 22.

27. "A Plan to Rescue the Louisiana Chemical Industry," *Chemical Week*, January 11, 1984, p. 10.

28. Story interview; "A Response to Safety and Health Allegations Released in Washington, D.C. by the Oil, Chemical, and Atomic Workers Union on March 6, 1986," June 11, 1986, p. 33, "OCAW and Local 4-620 BASF Proposed Exhibit 447" folder, box 19, IUP.

29. Dedeaux interview.

30. L. J. Story to D. J. Buchner, August 8, 1984, Story Papers.

31. B. F. Lessing to L. J. Story, August 2, 1984, Story Papers; Story interview.

32. "Locked-Out Louisiana Workers Testify against German Chemical Giant BASF before House Labor-Management Subcommittee," OCAW press release, March 30, 1988, "John Daigle" box, LUP.

33. L. J. Story to E. F. Gremillion, October 9, 1984, "Labor-Negotiations" folder, box 6, IUP.

34. L. J. Story to E. F. Gremillion, October 23, 1984, "Labor-Negotiations" folder, box 6, IUP; L. J. Story to E. F. Gremillion, November 14, 1984, "Labor-Negotiations" folder, box 6, IUP.

35. L. J. Story to E. F. Gremillion, October 23, 1984, "Labor-Negotiations" folder, box 6, IUP.

36. Edwin W. Edwards to Kernest J. Lanoux, August 9, 1984, "Corporate Campaign" folder, box 4, LUP; J. Bennett Johnston to Angela Gremillion, March 19, 1985, untitled folder, "Richardson" box, LUP. See also J. Bennett Johnston to Mary Schouest, January 11, 1985, untitled folder, "Richardson" box, LUP.

37. Mrs. Patsy Cudd to the Honorable Ronald Wilson Reagan, January 8, 1985, untitled folder, "Richardson" box, LUP; K. J. Lanoux to Edward [*sic*] Stenzel, August 19, 1984, untitled folder, "Richardson" box, LUP; Edwin L. Stenzel to Howard D. Samuel, September 10, 1984, "Correspondence BASF to OCAW" folder, box 2, IUP.

38. Notes of Meeting between Esnard Gremillion, Fred Loy, and Bob Gallant, September 18, 1987, "Religious Letters on BASF Lockout" folder, box 4, LUP.

39. Tim Talley, "End to BASF Lockout Termed Indefinite," *Baton Rouge Morning Advocate*, June 14, 1985, clipping in Clipping file, "Richardson" box, LUP.

40. Kettenacker, *Germany since 1945*, 96–7; Berghahn and Karsten, *Industrial Relations*, 214.

41. Berghahn, *Modern Germany*, 306.

42. Kettenacker, *Germany since 1945*, 96–7, Berghahn and Karsten, *Industrial Relations*, 38.

43. Kettenacker, *Germany since 1945*, 96–7 (quotation on p. 97).

44. "German CPI Union: Gains without Strife," *Chemical Week*, September 18, 1985, p. 40.

45. Berghahn, *Modern Germany*, 243; Berghahn and Karsten, *Industrial Relations*, 213; Leonard interview.

46. Leonard interview.

47. "Co-Conspirators of the Third Reich," n.d., Story Papers; Borkin, *The Crime and Punishment*; Schneider interview.

48. Rousselle interview; 'Shug' Allen to the Editor, *Gonzales Weekly*, September 21, 1984, clipping in "ULP15–CB-3195" folder, box 18, IUP; "Labor Related Incidents," 8/10/84–5/25/85, Story Papers.

49. Daigle, quoted in Tim Talley, "Union Leaflets Attack BASF Corp.," *Baton Rouge Morning Advocate*, January 10, 1985, clipping in "ULP15–CB-3195" folder, box 18, IUP; "Geismar, Louisiana OCAW and IUD Corporate Campaign Sequence of Events," n.d., Story Papers.

50. Undated photograph, Story Papers; Story interview.

51. Bernd Leibfried to Joseph M. Misbrener, July 15, 1985, Potential Exhibit 296, box 17, IUP; Telegram from I.G. Chemie dated June 28, 1985, to Joseph M. Misbrener, Potential Exhibit 296, box 17, IUP.

52. James A. Mattson to L. Calvin Moore, September 4, 1985, "I.G. Chemie" folder, box 13, IUP; Robert E. Wages to Werner Vitt, July 22, 1985, "I.G. Chemie" folder, box 13, IUP.

53. "Oil Workers to Go Abroad to Pursue Dispute against German Chemical Firm," *DLR*, January 17, 1985, p. A10, copy in "Labor-Negotiations" folder, box 6, IUP.

54. "'Coordinated Corporate Campaigns' and the Local Labor Dispute at BASF's Geismar Works Plant: A Backgrounder," March 6, 1986, p. 3, "Corporate Campaign Effectiveness" folder, box 2, IUP.

55. Transcript of *Locked Out!* video, October 7, 1988, pp. 7–8, "BASF v. OCAW—Correspondence No. 3" folder, box 12, IUP; Wages interview.

56. L. J. Story to E. F. Gremillion, February 26, 1985, "Labor-Negotiations" folder, box 6, IUP.

57. "Plant's Lockout Brings Tough Times," *New Orleans Times- Picayune*, April 22, 1985, p. A33.

Chapter 4. Bhopal on the Bayou?

1. Richard Leonard to Erik Lenar, September 13, 1985, "1985 Correspondence" folder, box 2, IUP.

2. Misbrener, quoted in "'Coordinated Corporate Campaigns' and the Local Labor Dispute at BASF's Geismar Works Plant: A Backgrounder," March 6, 1986, p. 7, "Corporate Campaign Effectiveness" folder, box 2, IUP.

3. "BASF Lockout of OCAW Local 4-620," May 1990, tape transcription, p. 16, box 8, IUP.

4. Juravich and Bronfenbrener, *Ravenswood*, 69–71; Hodges, "The Real Norma Rae," 262–8; Hodges, "J. P. Stevens and the Union," 61–3.

5. Juravich and Bronfenbrener, *Ravenswood*, 70–2.

6. "Victory at Memphis," *OCAW Reporter*, May-June 1985, pp. 14, 24; Richard Leonard to Joseph Misbrener, February 7, 1984, "Weekly Reports—Original" folder, "MAPCO No. 2" box, IUP.

7. Leonard interview.

8. A Concerned Shareholder to Mr. Habert and Mr. Cashman, March 5, 1985, "Letters—to Company and Shareholders" folder, "R. P. Scherer" box, IUP; Joseph M. Misbrener to R. P. Scherer, February 6, 1985, "Letters to Membership" folder, "R. P. Scherer" box, IUP.

9. "A Labor Union Wields 'PR' Weapons," *Chemical Week*, January 22, 1986, p. 57; "ABC 20/20," videotape, box 8, IUP, copy in author's possession.

10. Richard Leonard to E. J. Rousselle, July 9, 1985, "Strategy" folder, box 12, IUP.

11. "BASF Lockout of OCAW Local 4-620," tape transcription, p. 17, box 8, IUP.

12. Richard Leonard to Leslie Israel, September 11, 1985, "1985 Correspondence" folder, box 2, IUP.

13. Mazzochi interview; Leonard interview; Miller interview.

14. Miller interview.

15. Ibid.

16. Orr interview; Fontenot interview; Schneider interview.

17. Richard Leonard to Erik Lenar, September 13, 1985, "1985 Correspondence" folder, box 2, IUP.

18. "The Bhopal Disaster: How It Happened," *New York Times*, January 28, 1985, p. A1; "Carbide and India Strive for Bhopal Fund Accord," *New York Times*, November 19, 1987, p. D5; "Bhopal Payments by Union Carbide Set at $470 Million," *New York Times*, February 15, 1989, p. A1. For a detailed investigation into the accident, see Everest, *Behind the Poison Cloud*.

19. "Bridging the Bhopal Gap," *Chemical Week*, May 1, 1985, p. 3; "Bhopal: Legislative Fallout in the U.S.," *Chemical Week*, February 6, 1985, pp. 26–8.

20. "Bhopal: Legislative Fallout in the U.S.," *Chemical Week*, February 6, 1985, p. 26.

21. Ibid.

22. Stewart Diamond, "Credibility a Casualty in West Virginia," *New York Times*, August 18, 1985, sec. 4, p. 1. See also Thomas J. Lueck, "Chemical Industry Braces for Tougher Regulation," *New York Times*, August 15, 1985, pp. A1, A16.

23. "Oil Workers to Go Abroad to Pursue Dispute against German Chemical Firm," *DLR*, January 17, 1985, p. A9, copy in "Labor-Negotiations" folder, box 6, IUP.

24. Miller interview; Rousselle interview.

25. "Another Bhopal?" IUD Press Release, December 13, 1985, Story Papers.

26. Green Party Press Release, December 17, 1985, filed as "Potential Exhibit 359," box 17, IUP.

27. "Bhopal on the Bayou?" union flyer, untitled folder, untitled box, LUP.

28. "A Bhopal on the Bayou?" *New York Times,* January 5, 1986, sec. 3, p. 15; "Is It Safe?" *Wall Street Journal,* December 24, 1985, p. 1; Matt Shade to Robert G. Thoma, May 28, 1986, and Robert G. Thoma to Kenneth B. Noble, January 8, 1986, both in Story Papers.

29. Orr interview; "Take Down That Sign," *Baton Rouge Morning Advocate,* August 23, 1987, p. 10B.

30. Jenkins interview; Matt Shade to Robert G. Thoma, May 28, 1986, Story Papers.

31. Daisy G. Goodlaw to BASF Wyandotte Corporations, March 17, 1986; J. Carey Frederic to Les Story, December 17, 1985; Robert G. Thoma to Kenneth B. Noble, January 8, 1986, Story Papers.

32. Story interview.

33. Richard Leonard to Ron Stroman, February 26, 1988, "1988 Correspondence" folder, box 2, IUP.

34. OCAW Submissions of March 6, 1986, and February 28, 1988, both in "Conyers Study" folder, box 3, IUP (quotations on pp. 2, 3 of February 28 document).

35. OCAW Submission of February 28, 1988, pp. 1, 3, 4–5, "Conyers Study" folder, box 3, IUP (quotation on p. 5).

36. Quoted in "BASF's merits debated," *Terre Haute (Ind.) Tribune-Star,* May 8, 1988, clipping in "Terre Haute" folder, box 15, IUP.

37. Story interview; Jerry Blizin to Robert G. Thoma, November 6, 1986, Story Papers.

38. Lee Griffin to J. Bennett Johnston, January 16, 1986, Russell Long to A. J. De Marcay, February 7, 1986, Edwin W. Edwards to Les Story, March 27, 1987, Story Papers.

39. Leslie J. Story to the Honorable John Conyers, June 11, 1986, "OCAW and Local 4-620 BASF Proposed Exhibit 447" folder, box 19, IUP.

40. "A Response to Safety and Health Allegations Released in Washington, D.C., by the Oil, Chemical, and Atomic Workers Union on March 6, 1986," June 11, 1986, pp. ii, 2, 4, 28, 35, "OCAW and Local 4-620 BASF Proposed Exhibit 447" folder, box 19, IUP (quotations on pp. ii, 2, 35).

41. Ibid. (quotations on pp. 4–5, 46).

42. Story interview.

43. "JOP" to L. J. Story, June 6, 1986, Story Papers.

44. J. Platt to L. J. Story, n.d., Story Papers.

45. L. J. Story to Greg Landry, March 1, 1985, L. J. Story to Yank Kliebert, March 1, 1985, L. J. Story to Ernie Guitreau, March 1, 1985, Story Papers.

46. Guidry interview.

47. J. F. Hamilton to E. L. Stenzel, February 5, 1986, Story Papers. See also Mark L. Taylor to L. J. Story, March 11, 1985, Story Papers; Vern Bidwell to J. D. Martinez, February 17, 1986, Story Papers.

48. "Proud to be at B.W.C.," January 21, 1986, Story Papers.

49. John A. Berthelot to Juergen F. Strube, January 21, 1986, Story Papers.

50. William B. Little to Juergen F. Strube, January 29, 1986, Story Papers.

51. E. L. Stenzel to J. F. Hamilton, February 19, 1986, Story Papers.

52. "Locked-Out Workers Go Public to Pressure BASF," *Multinational Monitor*, March 15, 1986, p. 11.

53. Robert E. Wages to Clarke Ellis, May 1, 1985, "OECD Anti-Trust Complaint 3" folder, box 10, IUP; Walter B. Lockwood to Robert E. Wages, August 2, 1985, "OECD Anti-Trust Complaint 8" folder, box 11, IUP; Richard Leonard to Dan Berman, November 12, 1986, "1986 Correspondence" folder, box 2, IUP; Jenkins interview.

54. BASF Press Release, March 6, 1986, "Corporate Campaign Effectiveness" folder, box 2, IUP.

55. "'Coordinated Corporate Campaigns' and the Local Labor Dispute at BASF's Geismar Works Plant: A Backgrounder," p. 1, March 6, 1986, "Corporate Campaign Effectiveness" folder, box 2, IUP.

56. "BASF Charge against Union Rejected by NLRB," undated OCAW Press Release, in author's possession; Joseph M. Misbrener to Brothers and Sisters, January 6, 1987, "1987 Correspondence" folder, box 2, IUP.

Chapter 5. The Campaign Escalates

1. Hasten interview; Fontenot interview.

2. Schwab, *Deeper Shades of Green*, 217–23; Bullard, *Dumping in Dixie*, 65–9.

3. Schwab, *Deeper Shades of Green*, 223–9, 253–55 (quotation on p. 254).

4. Guste interview; Schwab, *Deeper Shades of Green*, 212.

5. Schwab, *Deeper Shades of Green*, 217–28 (quotation on p. 209).

6. Guste interview; Fontenot interview.

7. Susan Reed, "His Family Ravaged by Cancer, an Angry Louisiana Man Wages War on the Very Air That He Breathes," *People Weekly*, March 25, 1991, pp. 42, 43, 44, 48 (quotations on pp. 42, 43).

8. "All Clear for Seafood?" *New Orleans Times-Picayune*, August 15, 1980, sec. 1, p. 10.

9. "The Poisoned Land," *New Orleans Times-Picayune*, September 8–13, 1985, supplement, p. 1, clipping in "Locations/Louisiana" folder, box 5, IUP. See also "Prosperity in Paradise? Louisiana's Chemical Legacy," *Baton Rouge Morning Advocate*, April 25, 1985, supplement, clipping in "Locations/Louisiana" folder, box 5, IUP.

10. Favorite interview.

11. Ibid.

12. Susan Reed, "His Family Ravaged by Cancer, an Angry Louisiana Man Wages War on the Very Air That He Breathes," *People Weekly*, March 25, 1991, pp. 42, 43, 44, 48 (quotation on p. 44); Fontenot interview.

13. Favorite interview.

14. Ibid. For a detailed account of the 1965 Selma to Montgomery march, see Garrow, *Protest at Selma*, 73–90, 116–7.

15. Amos J. Favorite statement, n.d., "Groups/Enviro/APRATP" folder, box 5, IUP.

16. "Comments on the Proposed New Class I Injection by BASF Corporation, DNR Docket No. UIC 87–31, Submitted by Ascension Parish Residents against Toxic Pollution, Geismar, Louisiana," November 13, 1987, "BASF/Injection Wells" folder, box 4, IUP; Miller interview.

17. "Statement from the Ascension Parish Residents against Toxic Pollution," September 24, 1986, "Groups/Enviro/APRATP" folder, box 5, IUP.

18. Richard Miller to Ross Vincent, December 7, 1986, "BASF Incinerator Geismar" folder, box 4, IUP.

19. Arnold interview; Favorite interview.

20. Hawkins interview.

21. Amos J. Favorite to H. S. Simon, February 17, 1987, "Groups/Enviro/APRATP" folder, box 5, IUP.

22. C. D. Porter to Amos J. Favorite, February 19, 1987, Dan Fisher to Amos J. Favorite, February 19, 1987, L. J. Story to Amos Favorite, December 22, 1986, "Groups/Enviro/APRATP" folder, box 5, IUP.

23. Malek-Wiley interview. For information on the Shell strike, see Gordon, "Shell No!"

24. Malek-Wiley interview.

25."Superfund Poses Major Problems for Industry," *Industry Week*, September 21, 1987, clipping in "Media/Clips/Stats" folder, box 15, IUP; Fontenot interview.

26. "The Routine and Accidental Release of Chemicals into the Mississippi River by Fifteen Plants Located in the Geismar, Saint Gabriel, and Plaquemine, Louisiana Area," January 13, 1986, "Sierra Club—Water" folder, box 15, IUP; "A Preliminary Report to the Louisiana House Natural Resources Subcommittee on Oversight Concerning the Annual Dumping of Almost 200 Million Pounds of Toxic Chemicals into the Air in the Geismar and Saint Gabriel Communities by Eighteen Petrochemical Plants and Numerous Accidental Toxic Chemical Releases during the First Nine Months of 1986," October 21, 1986, "Sierra Club—Air" folder, box 15, IUP; Schwab, *Deeper Shades of Green*, 236.

27. Transcript of *Locked Out!* video, October 7, 1988, p. 18, "BASF v. OCAW Correspondence No. 3" folder, box 12, IUP.

28. "Narrative," p. 1, "LEAN Narrative 1990" folder, LEAN Papers.

29. "Articles of Incorporation: Louisiana Environmental Action Network," January 5, 1987, p. 1, "BASF Intervenor Suit" folder, LEAN Papers.

30. Orr interview.

31. Untitled document, pp. 5–6, "LEAN Narrative 1990" folder, LEAN Papers; Orr interview.

32. "Original Brief of the Appellants-Plaintiffs Sierra Club and Ascension Parish Residents against Toxic Pollution," October 26, 1987, p. 4, "BASF Court Case" folder, LEAN Papers.

33. "Sierra Club Suit Challenges DEQ's Penalty Policies," n.d., "Intervenor Suit BASF" folder, LEAN Papers.

34. Ibid.

35. Richard Miller to LEAN executive board, July 22, 1987, "Intervenor Suit BASF" folder, LEAN Papers; Notice of Judgement, *BASF Corporation v. APRATP et al.*, December 20, 1988, "Intervenor Suit BASF" folder, LEAN Papers.

36. Miller interview; Fontenot interview.

37. "Original Brief of Appellee BASF Corporation," May 21, 1987, pp. 2–6, "BASF Court Case" folder, LEAN Papers (quotations on pp. 2, 6).

38. Notice of Judgement, *BASF Corporation v. APRATP et al.*, December 20, 1988, "Intervenor Suit BASF" folder, LEAN Papers (quotation on p. 17); Grant Proposal, October 5, 1991, "John Daigle" box, LUP.

39. Hulsberg, *The German Greens*, 100, 103, 199–200, 210.

40. Richard Leonard to Willi Hoss, January 15, 1988, "1988 Correspondence" folder, box 2, IUP; Richard Leonard to Dirk Oblong, February 16, 1988, "1988 Correspondence" folder, box 2, IUP.

41. "Geismar, Louisiana, OCAW and IUD Corporate Campaign Sequence of Events," n.d., Story Papers; Suhr, quoted in "Locked-Out Workers Go Public to Pressure BASF," *Multinational Monitor*, March 15, 1986, p. 11.

42. Tim Talley, "Germans Call Rejection of Second Bid to Visit Plant 'Scandalous,'" *Baton Rouge State Times*, April 11, 1986, clipping in "Media/Clips/Statistics" folder, box 15, IUP; Miller interview; Steimel, quoted in Tim Talley, "German Officials Refused Entrance to BASF Plant," *Baton Rouge Morning Advocate*, April 11, 1986, clipping in "Media/Clips/Statistics" folder, box 15, IUP.

43. Richard Leonard, address, "Annual Meeting 1986" folder, box 1, IUP; Leonard interview.

44. Albers, quoted in L. J. Story to All Geismar Employees, June 26, 1986, "Annual Meeting 1986" folder, box 1, IUP.

45. L. J. Story to All Geismar Employees, June 26, 1986, "Annual Meeting 1986" folder, box 1, IUP; "Summary Evaluation and Possible Directions for the Campaign Against BASF," September 1986, "Strategy" folder, box 12, IUP.

46. Hulsberg, *The German Greens*, 160, 171 (quotation on p. 171).

47. Joseph B. Uehlein to Richard Leonard, December 19, 1986, Lane Kirkland to Ernst Breit, November 11, 1986, "Unions—IUD" folder, box 13, IUP.

48. Richard Leonard to Joe Uehlein, December 29, 1986, "Unions—IUD" folder, box 13, IUP.

49. Richard Leonard to Joseph Misbrener, December 29, 1986, "I.G. Chemie" folder, box 13, IUP; "Aussperrung durch amerikanisches BASF-Tochterunterneh-

men," *Politik und Zeitgeschehen,* undated clipping and translation in "I.G. Chemie" folder, box 13, IUP; Wages interview.

50. Angelo Bartolo to Esnard Gremillion, February 5, 1986, Jose Drummond to Richard Leonard, September 17, 1986, "Unions—General" folder, box 13, IUP.

51. ICEF Leaflet, "ICEF" folder, box 13, IUP; Vic Thorpe to Richard Leonard, March 2, 1988; Richard Leonard to Pekka Aro, March 17, 1986, "Unions—ICEF" folder, box 13, IUP.

52. "A Call to BASF and Hitachi to End Computer Sales to South Africa," December 1986, D. Heckle to Timothy H. Smith, December 12, 1986, Hirokichi Yoshiyama to Timothy H. Smith, December 26, 1986, "South Africa—Reports and Correspondence" folder, box 12, IUP.

53. "Jesse Jackson, Labor and Anti-Apartheid Leaders Call on BASF and Hitachi to End Computer Sales to South Africa," December 11, 1986, "South Africa—Reports and Correspondence" folder, box 12, IUP; "Like a Brush Fire" (translation), *Der Spiegel,* March 2, 1987, "South Africa No. 1" folder, box 12, IUP.

54. Dick Leonard to Richard Leonard, March 25, 1988, "South Africa—Reports and Correspondence" folder, box 12, IUP; John E. Mullane to Robert G. Thoma, July 9, 1987, Story Papers.

55. Tim Talley, "Jesse Jackson Condemns Lockout at BASF Plant," *Baton Rouge Morning Advocate,* January 22, 1987, clipping in "Groups/Church/Jackson" folder, box 5, IUP.

56. John R. Daigle et al. to Reverend Jackson, n.d., Leslie S. Vann to Rev. Jesse Jackson, May 5, 1987, "Groups/Church/Jackson" folder, box 5, IUP.

57. Joseph M. Misbrener to Reverend Jesse Jackson, February 2, 1987, "Groups/Church/Jackson" folder, box 5, IUP; Bill McMahon, "Jackson Tells Union Environment Needs Protection," *Baton Rouge State Times,* September 22, 1988, clipping in "Media/Clips/Stats" folder, box 15, IUP.

58. "A Labor Union Wields 'PR' Weapons," *Chemical Week,* January 22, 1986, p. 57; Story interview; Donaldson interview.

59. "Geismar, Louisiana, OCAW and IUD Corporate Campaign Sequence of Events," n.d., Rafael Bermudez to Les Story and Richard Donaldson, November 19, 1986, Story Papers.

60. "Responses to the Torpedo," n.d., Story Papers.

61. L. J. Story to All Salaried Employees, July 16, 1984, "Correspondence—BASF to OCAW" folder, box 2, IUP; Richard Leonard to Reverend George Ogle, March 27, 1987, "1987 Correspondence" folder, box 2, IUP; Hawkins interview.

62. L. J. Story to E. F. Gremillion, August 25, 1986, "Labor-Negotiations" folder, box 6, IUP.

63. "'Coordinated Corporate Campaigns' and the Local Labor Dispute at BASF's Geismar Works Plant: A Backgrounder," March 6, 1986, pp. 1–2, 5–6, "Corporate Campaign Effectiveness" folder, box 2, IUP; Rousselle interview.

64. William L. Jenkins to Ernest J. Rousselle, September 17, 1984, "Labor-Negotiations" folder, box 6, IUP.

65. Larry E. Kelly to L. J. Story, February 26, 1986, Story Papers; Harvey interview.

66. Joseph M. Misbrener to All Local Unions, April 16, 1987, E. F. Gremillion to Brothers and Sisters, April 1987, "John Daigle" box, LUP; Schwab, *Deeper Shades of Green*, 239; Joseph M. Misbrener to All Local Union and Council Secretaries, September 18, 1986, untitled folder, "Richardson" box, LUP.

67. "Summary Evaluation and Possible Directions for the Campaign against BASF," n.d., "Strategy" folder, box 12, IUP. Although undated, the events described in this detailed document make it clear that it was written toward the end of 1986. For similar correspondence, see Richard Leonard to Kate Karam, October 24, 1986, "1986 Correspondence" folder, box 2, IUP.

Chapter 6. Breakthrough

1. Joseph M. Misbrener to Brothers and Sisters, January 6, 1987, Richard Leonard to David Weiner, February 10, 1987, "1987 Correspondence" folder, box 2, IUP.

2. Richard Leonard to Tina Stadlmayer, May 19, 1987, Richard Leonard to Willi Hoss, April 13, 1987, Richard Leonard to Patrick J. Tully, May 12, 1987, "1987 Correspondence" folder, box 2, IUP.

3. "BASF 'Locked In' at New Jersey Headquarters," OCAW Press Release, n.d., "Corporate Campaign" folder, box 4, LUP.

4. "Convictions Upheld for Eighteen Protesters," *Newark (N.J.) Star-Ledger*, December 30, 1987, clipping in "John Daigle" box, LUP.

5. Richard Leonard to Brothers, Sisters, and Friends, June 16, 1987, "1987 Correspondence" folder, box 2, IUP. For press coverage, see, for example, "Transcript of Press Conference," June 24, 1987, "1987 Correspondence" folder, box 2, IUP; "Protesters Seized in Picketing of BASF Corp. Headquarters," *New York Times*, June 4, 1987, p. B3.

6. J. E. Blouin to All Parsippany Employees, June 1, 1987, Memo to all Parsippany Employees, June 3, 1987, Story Papers.

7. Story interview.

8. Les Story, address to Rotary Club, September 24, 1986, Story Papers.

9. Richard Leonard to Stephen Lewis, July 20, 1987, "1987 Correspondence" folder, box 2, IUP; Henry Kramer, letter to author, December 11, 2000, copy in author's possession; Story interview.

10. "BASF Threatens OCAW Leaders with Fines, Jail Terms," March 18, 1988, Robert E. Wages to All Chemical Industry Managers Who Are Concerned with BASF, February 17, 1988, "BASF Press Releases" folder, box 3, LUP.

11. "Who's New on Labor's 'Dishonor Roll,'" *New York Times*, January 5, 1986, sec. 3, p. 15. See also "The Dishonor Roll," *Wall Street Journal*, October 29, 1985, p. 1.

12. "Response of the BASF Board of Directors to the Questions from Shareholder Representatives Richard Leonard and Steven Cate," June 26, 1987, "Annual Meeting 1987" folder, box 1, IUP.

13. Hoss, quoted in "Press Release No. 522/87," June 26, 1987, "Annual Meeting 1987" folder, box 1, IUP. See also "Chemical Industry Employees Were Not All Satisfied" (translation), *Der Mannheimer Morgen Post*, June 26, 1987, translation in "Annual Meeting 1987" folder, box 1, IUP.

14. "Do You Want Another Chemical Waste Incinerator in Our Parish?" APRATP leaflet, September 1987, "APRATP Fact Sheet: BASF Corporation's Proposed Amines," July 6, 1988, "BASF General" folder, LEAN Papers.

15. Darryl A. Stevens to Mr. Koury, March 3, 1987, "BASF Incinerator Geismar" folder, box 4, IUP.

16. L. J. Story to Gary L. Keyser, August 31, 1987, "BASF Incinerator Geismar" folder, box 4, IUP.

17. "BASF Corporation Response to Ascension Parish Residents against Toxic Pollution Comments on the Proposed BASF Solid Waste Incinerator," June 16, 1987, p. 2, "BASF Incinerator Geismar" folder, box 4, IUP.

18. Richard Miller to Richard J. Cook, January 8, 1988, "BASF Incinerator Geismar" folder, box 4, IUP.

19. "Message from the General Manager," April 8, 1988, "BASF Incinerator Geismar" folder, box 4, IUP.

20. Bob Anderson, "Firm Plans Incinerator," *Baton Rouge Morning Advocate*, November 8, 1985, clipping in "BASF Incinerator Geismar" folder, box 4, IUP.

21. "Parish to Enjoy Additional $620,000 in Revenues," *New Orleans Daily Record*, December 30, 1987, clipping in "Clips/Media/Stats" folder, box 15, IUP.

22. Miller interview; William J. Guste to Sandra Robinson, August 19, 1987, "Locations/St. Gabriel" folder, box 5, IUP.

23. Sharon Donovan, "La. Women Fear Plants, Birth Problems Linked," *USA Today*, August 21–23, 1987, clipping in "Conyers Study" folder, box 3, IUP; David Maraniss and Michael Weisskopf, "Jobs and Illness in Petrochemical Corridor," *Washington Post*, December 22, 1987, clipping in "Media/Clips/Statistics" folder, box 15, IUP; Richard Miller to Steve Davis, February 19, 1988, "Locations/St. Gabriel" folder, box 5, IUP.

24. Jeff D. Opdyke, "Coping with Chemicals: The St. Gabriel Story," *Louisiana State University Daily Reveille*, October 14, 1987, p. 11; David Maraniss and Michael Weisskopf, "Jobs and Illness in Petrochemical Corridor," *Washington Post*, December 22, 1987, clipping in "Media/Clips/Statistics" folder, box 15, IUP.

25. Marylee M. Orr to Katherine Graham, November 16, 1989, "1989 LEAN Outgoing" folder, LEAN Papers.

26. William J. Guste to Sandra Robinson, August 19, 1987, "Locations/St. Gabriel" folder, box 5, IUP.

27. "Final Report: St. Gabriel Miscarriage Investigation, East Bank of Iberville Parish, Louisiana," September 27, 1989, p. 1, LUP.

28. Malek-Wiley interview; John Daigle, "Comments on Ciba Geigy's Hazardous Waste Incinerator Permit Application, St. Gabriel," May 24, 1989, John Daigle box, LUP.

29. Schwab, *Deeper Shades of Green*, 237–8; Miller interview; Malek-Wiley interview.

30. David Maraniss and Michael Weisskopf, "Jobs and Illness in Petrochemical Corridor," *Washington Post*, December 22, 1987, clipping in "Media/Clips/Statistics" folder, box 15, IUP; Susan Reed, "His Family Ravaged by Cancer, an Angry Louisiana Man Wages War on the Very Air That He Breathes," *People Weekly*, March 25, 1991, pp. 42, 43, 44, 48; Miller interview.

31. Patterson, *The Dread Disease*, vii.

32. Carson, *Silent Spring*, 178–99, esp. 186–7; Patterson, *The Dread Disease*, 282–3; Nader, Brownstein, and Richard, *Who's Poisoning America*, 10–2. For a detailed examination of the fight to clean up the Love Canal site, see Brown, *Laying Waste*, xi–96; Epstein, Brown, and Pope, *Hazardous Waste*, 89–132.

33. "The Poisoned Land," *New Orleans Times-Picayune*, September 8–13, 1985, p. 5, "Prosperity in Paradise?: Louisiana's Chemical Legacy," *Baton Rouge Morning Advocate*, April 25, 1985, supplement, pp. 1, 3, 7, clippings in "Locations/Louisiana" folder, box 5, IUP.

34. Story interview; Steimel interview.

35. Orr interview; Richard Donaldson, letter to author, November 26, 2000, copy in author's possession; W. C. Moran to Paul Templet, October 19, 1988, Miller Papers.

36. *Louisiana Industry Environmental Alert* 4, no. 5 (May 1989): 1, copy in "News Articles" folder, "BASF Lockout" box, LUP; Duncan, *Goodbye Green*, 32.

37. "1000 March on Third Anniversary of Lockout," OCAW Press Release, n.d., "Corporate Campaign" folder, box 4, LUP.

38. "BASF 'Locked In' at New Jersey Headquarters," OCAW Press Release, n.d., "Corporate Campaign" folder, box 4, LUP; George E. Ogle to Les Story, March 30, 1987, Story Papers.

39. Joseph M. Misbrener to Howard Samuels, July 23, 1987, "1987 Correspondence" folder, box 2, IUP; Leonard interview; Rousselle interview.

40. Stanley Joseph Ott, "1987 Labor Day Homily," September 7, 1987, "Gov. Edwards Letter and Resolutions" folder, box 4, LUP; Stanley Joseph Ott to Esnard Gremillion, June 23, 1987, "Religious Letters on BASF Lockout" folder, box 4, LUP.

41. Rev. George F. Lundy, S.J., and William Temmink, "A Battle on the Mississippi," *Blueprint for Social Justice*, April 1987, pp. 1–6 (quotations on pp. 1, 4, 5, 6).

42. Rev. George F. Lundy to Rev. Alfred C. Kramer, April 1, 1985, "Women's Support Group" folder, box 4, LUP; Fara Impastato to Richard Miller, November 4, 1988, "Richardson" box, LUP; Transcript of *Locked Out!* video, October 7, 1988, p. 7, "BASF v. OCAW Correspondence No. 3" folder, box 12, IUP.

43. E. F. Gremillion to the Most Reverend Stanley J. Ott, July 1, 1987, "Reli-

gious Letters on BASF Lockout" folder, box 4, LUP; Hawkins interview. Marion "Putsy" Braud expressed similar views as Hawkins (Braud interview).

44. "Death Takes Three Members of Local 4-620," OCAW Press Release, n.d., "Corporate Campaign" folder, Box 4, LUP; Arnold interview; *Baton Rouge Morning Advocate*, July 2, 1987, annotated clipping, in "BASF Lockout" box, LUP.

45. "Death Takes Three Members of Local 4-620," OCAW Press Release, n.d., "Corporate Campaign" folder, box 4, LUP.

46. Miller interview; Leonard interview.

47. Fink interview.

48. Braud interview.

49. "Citizen Fact-Finding Proposal Gains Momentum," OCAW Press Release, n.d., "Corporate Campaign" folder, box 4, LUP.

50. L. J. Story to the Honorable Edwin W. Edwards, July 24, 1987, "Correspondence—BASF" folder, box 2, IUP.

51. Edwin W. Edwards to Don F. Perkins, August 11, 1987, "Gov. Edwards Letter and Resolutions" folder, box 4, LUP.

52. John E. Mullane to Robert G. Thoma, July 9, 1987, Story Papers.

53. Rafael Bermudez to Les Story et al., May 18, 1987, Story Papers; James E. Fitzmorris to Leslie J. Story, June 9, 1987, Story Papers.

54. L. J. Story to Mary Lee Orr, September 4, 1987, "Corporate Campaign Effectiveness" folder, box 2, IUP.

55. Leslie Story, "Not Just a 'Good Guy/Bad Guy' Issue," *Shreveport Times*, February 1, 1987, p. 8B.

56. R. W. Lutz to Therese Murtagh, November 10, 1987, untitled folder, box 2, LUP.

57. Amos Favorite to Paul Miller, September 27, 1987, untitled folder, box 2, LUP; Favorite interview.

58. Transcript of *Locked Out!* video, October 7, 1988, p. 17, "BASF v. OCAW Correspondence No. 3" folder, box 12, IUP.

59. Dick Leonard to James Hughes, September 3, 1987, IUP.

60. Dick Leonard to John McKendree, February 9, 1987, box 2, IUP; Miller interview.

61. Notes of Meeting, September 18, 1987, "Religious Letters on BASF Lockout" folder, box 4, LUP.

62. Company Memorandum, September 1987, Story Papers; D. J. Buchner to J. E. Blouin, August 3, 1987, Story Papers.

63. John E. Mullane to Robert G. Thoma, July 9, 1987, Story Papers.

64. Company Memorandum, September 1987, Story Papers.

65. Donaldson interview.

66. Berghahn and Karsten, *Industrial Relations*, 102.

67. Ibid., 89.

68. Ibid., 89, 96.

69. Jenkins interview.

70. Pederson, ed., *International Directory*, 50.

71. "Transcript of Interview between L. J. Story and Hermann Vinke," n.d., Story Papers.

72. Nicola Palmieri-Egger, letter to author, January 12, 2001, copy in author's possession. Other attempts to secure interviews with German executives were unsuccessful, with current BASF executives insisting that German management was not involved in the lockout. In addition, former BASF chairman Hans Albers died before I carried out the research for this project. John Kirkman, letter to author, February 2, 2001, copy in author's possession.

73. Nicola W. Palmieri-Egger to D. J. Buchner, October 8, 1986, D. J. Buchner to Nicola W. Palmieri-Egger, November 10, 1986, Story Papers.

74. Henry Kramer, letters to author, December 11, 2000, and February 6, 2001, copies in author's possession; Jenkins interview.

75. Story interview; Donaldson interview.

76. Story interview; Gilin interview.

77. F. R. Donaldson, notice, November 12, 1987, "Labor-Negotiations" folder, box 6, IUP; *DLR*, November 9, 1987, clipping in "Media/Clips/Stats" folder, box 15, IUP.

78. F. R. Donaldson to Supervisors/Managers of Union Represented Employees, September 9, 1988, F. R. Donaldson to W. C. Moran, December 13, 1988, "Incoming Correspondence" folder, "Richardson" box, LUP.

Chapter 7. It Ain't Over

1. Richard Leonard to Michael Braungart, May 6, 1988, "1988 Correspondence" folder, box 2, IUP.

2. "'End' to Lockout Declared 'Sham' by Local 4-620," OCAW Press Release, "Corporate Campaign" folder, box 4, LUP.

3. Nicola W. Palmieri-Egger to BASF Plants, February 12, 1988, "ULP 15–CA-10494" folder, box 19, IUP.

4. Schneider interview.

5. Hawkins interview.

6. Fink interview; Donaldson interview.

7. Grievance of Samuel E. Daigle et al., November 1988, "Grievances 1988" folder, "Richardson" box, LUP; Grievance of M. D. LeBlanc et al., February 9, 1989, Grievance of Kenneth Boudreaux, June 9, 1989, "Grievances 1989" folder, "Richardson" box, LUP.

8. Thomas W. Budd to E. J. Rousselle, January 8, 1988, "Labor-Negotiations" folder, box 6, IUP. See also Thomas W. Budd to E. J. Rousselle, April 18, 1988, "Labor-Negotiations" folder, box 6, IUP.

9. Schneider interview; Fink interview.

10. "Locked-Out Louisiana Workers Testify against German Chemical Giant

BASF before House Labor-Management Subcommittee," March 30, 1988, "John Daigle" box, LUP; "Filibusters Have Thwarted Labor Law Reform over the Years," *Paperworker*, September 1994, p. 8.

11. Badger, "When I Took the Oath," 15.

12. Ibid.

13. Richard Leonard to Joe Uehlein, April 5, 1988, "1988 Correspondence" folder, box 2, IUP.

14. Roemer interview.

15. Schwab, *Deeper Shades of Green*, 273–8 (statistic from pp. 273–4); Templet interview.

16. Steimel interview.

17. APRATP Press Release, n.d., "BASF/Injection Wells" folder, box 4, IUP; "Comments on the Proposed New Class I Injection by BASF Corporation, DNR Docket No. UIC 87–31, Submitted by Ascension Parish Residents against Toxic Pollution, Geismar, Louisiana," November 13, 1987, "BASF/Injection Wells" folder, box 4, IUP.

18. APRATP Press Release, February 25, 1988, "BASF Injection Wells" folder, box 4, IUP.

19. "BASF Corporation's Motion to Be Accorded 'Party' Status," November 12, 1987, "BASF Injection Wells" folder, box 4, IUP (quotation on p. 11).

20. "BASF Targeted by Citizens, Environmentalists," n.d., "BASF Press Releases" folder, box 3, LUP.

21. Bob Anderson, "Judge Orders DEQ to Draw Up Regulations," *Baton Rouge Morning Advocate*, March 5, 1988, clipping in "BASF/Injection Wells" folder, box 4, IUP; Bob Anderson, "DEQ to Halt Waste-Well Certifications," *Baton Rouge Morning Advocate*, March 4, 1988, p. 2B.

22. "BASF Targeted by Citizens, Environmentalists," n.d., "BASF Press Releases" folder, box 3, LUP.

23. H. C. Stafford to Fellow Employees, June 21, 1990, "BASF Injection Wells" folder, box 4, IUP.

24. Richard Miller to Jane Johnson et al., December 26, 1987, Richard Miller to James Welsh, March 25, 1989, both in "BASF Injection Wells" folder, box 4, IUP.

25. Richard Miller to Bob Cunningham, October 31, 1988, Miller Papers.

26. Miller's correspondence detailed in W. C. Moran to Paul Templet, October 19, 1988, Miller Papers.

27. Templet interview.

28. Templet, quoted in Frances Frank Marcus, "Labor Dispute in Louisiana Ends with Ecological Gain," *New York Times*, January 3, 1990, clipping in "Media/Clips/Stats" folder, box 15, IUP.

29. Roemer interview.

30. W. C. Moran to F. J. Federico et al., December 7, 1988, Story Papers.

31. W. C. Moran to Paul Templet, October 19, 1988, Miller Papers.

32. Richard Leonard to Carol Fagen, May 12, 1988, "1988 Correspondence" folder, box 2, IUP; Story interview.

33. Miller, quoted in Jim Ross, "Union Rep Tells Group to Question BASF Plans," *Huntington (W.Va.) Herald-Dispatch*, November 18, 1988, clipping in "Haverhill Area" folder, box 15, IUP. See also Susan Crittenden, "Waste Plans Divide Vigo County," *Indianapolis (Ind.) Star*, November 13, 1988, clipping in "Terre Haute" folder, box 15, IUP.

34. Jim Ross, "Union Rep Tells Group to Question BASF Plans," *Huntington (W.Va.) Herald-Dispatch*, November 18, 1988, Bill Thomas, "Union Rep Assails BASF, Local Plans," *Ironton (Ohio) Tribune*, vol. 141, no. 528, Leigh Stone, "PR Man Disputes Union Statements," *Portsmouth (Ohio) Daily Times*, November 18, 1988, clippings in "Haverhill Area" folder, box 15, IUP.

35. Larry Thomas, "BASF struck by Terre Haute 'Snowball,'" *Evansville (Ind.) Courier*, January 11, 1989, clipping in "Evansville Master" folder, box 15, IUP.

36. OCAW vs. BASF," *Postmouth (Ohio) Times*, December 20, 1988, clipping in "Haverhill Area" folder, box 15, IUP.

37. "Consider BASF's Past in Making a Decision," *Huntington (W.Va.) Herald-Dispatch*, January 22, 1989, "BASF Plant? No Way," *Huntington (W.Va.) Herald-Dispatch*, March 13, 1989, clippings in "Haverhill Area" folder, box 15, IUP.

38. Larry Thomas and John Reiter, "Profit, Turmoil Bubble in BASF's Caldron," *Evansville (Ind.) Courier*, March 27, 1988, clipping in "Evansville Master" folder, box 15, IUP.

39. "BASF Spells Jobs," company flyer in "Terre Haute" folder, box 15, IUP.

40. "BASF Reps Pleased with Response to Local Mailings," *Terre Haute (Ind.) Tribune-Star*, November 27, 1988, George W. Wardell, "Chalos Questions Motives of Vocal BASF Opponents," *Terre Haute (Ind.) Tribune-Star*, November 15, 1988, clippings in "Terre Haute" folder, box 15, IUP.

41. G. Sam Piatt, "Emotions Run High at Meeting about BASF Proposal," *Ashland (Ky.) Daily Independent*, February 2, 1989, clipping in "Haverhill Area Master" folder, box 15, IUP.

42. Wells E. Peach to Tri-State Environmental Group, February 4, 1989, "Huntington West Virginia—BASF Haverhill, Ohio—Proposed Site" folder, box 6, LUP.

43. Richard Miller to Richard Leonard, July 30, 1989, "Huntington West Virginia—BASF Haverhill, Ohio—Proposed Site" folder, box 6, LUP.

44. G. Sam Piatt, "Union Man against BASF in Area," *Ashland (Ky.) Independent*, November 18, 1988, clipping in "Haverhill Area" folder, box 15, IUP.

45. "Up from the Chemical Plantation," p. 71, "Media/Interview/Leonard/Benman" folder, box 9, IUP.

46. Doug Sword, "Skeptic: BASF May Have Taken 'Graceful Way Out,'" *Evansville (Ind.) Courier*, September 10, 1988, clipping in "Evansville Master" folder, box 15, IUP.

47. Jim Ross, "Haverhill Nearly Eliminated as Site for BASF Installation,"

Huntington (W.Va.) Herald Dispatch, April 8, 1989, clipping in "Haverhill Area" folder, box 15, IUP. See also G. Sam Piatt, "Soil Test Results Jeopardize Plans for BASF Plant," *Ashland (Ky.) Daily Independent,* February 1, 1989, clipping in "Haverhill Area" folder, box 15, IUP.

48. "Court Sends BASF Back to Square One," *Sullivan (Ind.) Daily Times,* July 3, 1989, clipping in "Terre Haute" folder, box 15, IUP.

49. "ILO Governing Body Critical of Chemical Giant BASF in 48–Month Lockout," June 1, 1988, "BASF Press Releases" folder, box 3, LUP.

50. "ILO Governing Body Critical of Chemical Giant BASF in 48–Month Lockout," June 1, 1988, "BASF Press Releases" folder, box 3, LUP; "BASF 'Lockout' Criticized by ILO, Cooperation of U.S. Sought," *Chemical Marketing Reporter,* June 6, 1988, clipping in "Clips/Media/Stats" folder, box 15, IUP.

51. Joseph M. Misbrener to Hermann Rappe, July 5, 1988, "1988 Correspondence" folder, box 2, IUP.

52. Joseph M. Misbrener to Hermann Rappe, September 2, 1988, "1988 Correspondence" folder, box 2, IUP.

53. "Summary Evaluation and Possible Directions for the Campaign against BASF," September 1986, "Strategy" folder, box 12, IUP.

54. Richard Leonard to Esnard Gremillion et al., September 8, 1988, "1988 Correspondence" folder, box 2, IUP.

55. Richard Leonard to Kent Curtis, February 18, 1988, "1988 Correspondence" folder, box 2, IUP.

56. Darryl Malek-Wiley, "The Great Louisiana Toxic March," *Blueprint for Social Justice,* October 1988, pp. 1–8, copy in "Great Louisiana Toxic March" folder, box 5, IUP; "The Great Louisiana Toxic March," n.d., "Great Louisiana Toxic March" folder, box 5, IUP.

57. Malek-Wiley interview.

58. Garrow, *Bearing the Cross,* 479–80; Garrow, *Protest at Selma,* 66–89; Williams, *Eyes on the Prize,* 164–205; Salmond, *"My Mind Set on Freedom,"* 129–33.

59. "Great Louisiana Toxics March: A Seventy-Mile March through Louisiana's Cancer Alley," flyer in "Toxic March" folder, untitled box, LUP.

60. "Join Us for the Great Louisiana Toxic March and Help Stop Toxic Exposures to Louisiana Workers!!!!," "Great Louisiana Toxic March" folder, box 5, IUP.

61. "Marchers Protesting Pollution," *Lafayette Advertiser,* November 12, 1988, clipping in "Great Louisiana Toxic March" folder, box 5, IUP.

62. "A Walk for Toxic Justice," *Environmental Action,* January-February 1989, p. 10, clipping in "Great Louisiana Toxic March" folder, box 5, IUP.

63. "Participants Mourn Passage of a Black Community," *White Castle Times,* November 16, 1988, clipping in "Toxic March" folder, untitled box, LUP.

64. "A Walk for Toxic Justice," *Environmental Action,* January-February 1989, p. 10, clipping in "Great Louisiana Toxic March" folder, box 5, IUP; "Marchers Pass

through Ascension Parish," *Gonzales (La.) Weekly*, November 18, 1988, clipping in "Toxic March" folder, untitled box, LUP.

65. "Sheriff Confronts Marchers over Fee for Police Escort," *Shreveport Times*, November 19, 1988, clipping in "Toxic March" folder, untitled box, LUP. During the civil rights movement, protesters often prayed in front of law enforcement officials who were opposing them. See, for example, Williams, *Eyes on the Prize*, 175, 274.

66. "A *Great* Great Toxics March," *Delta Sierran*, January 1989, pp. 1, 4; "Protesters Finish Off March," *Thibodaux (La.) Comet*, November 21, 1988, clipping in "Toxic March" folder, untitled box, LUP.

67. "Protesters End Bridge Holdout," *Thibodaux (La.) Comet*, November 17, 1988, clipping in "Toxic March" folder, untitled box, LUP.

68. "Greenpeace Tries to Block Discharge," *Baton Rouge Morning Advocate*, November 4, 1988, clipping in "Toxic March" folder, untitled box, LUP; "Environmentalists Split on Greenpeace Tactics," *Baton Rouge Morning Advocate*, November 4, 1988, clipping in "Toxic March" folder, untitled box, LUP.

69. "Environmentalists Plan March, Rally," *Louisiana State University Reveille*, November 2, 1988, clipping in "Toxic March" folder, untitled box, LUP; "Protesters Finish Off March," *Thibodaux (La.) Comet*, November 21, 1988, clipping in "Toxic March" folder, untitled box, LUP.

70. See, for example, "Tactics of Alarm Won't Give Answers," *Baton Rouge State-Times*, October 31, 1988, "Black Man Like Me Will Die of Cancer: The Great LA Toxics March to Clean Up Cancer Alley," *Monroe (La.) Dispatch*, November 3, 1988, clippings in "Toxic March" folder, untitled box, LUP.

71. "'Cancer Alley Is a Heavy-Duty Match for All the Positives," *La Place (La.) L'Observateur*, November 3, 1988, clipping in "Great Louisiana Toxic March" folder, box 5, IUP. For a sample of views, see "If You Breathe, Take a Look at It," *Baton Rouge Morning Advocate*, November 26, 1988, clipping in "Toxic March" folder, untitled box, LUP.

72. Malek-Wiley interview.

73. "Sheriff Confronts Marchers over Fee for Police Escort," *Shreveport Times*, November 19, 1988, clipping in "Toxic March" folder, LUP; "Historic Toxic March Brings Attention to Cancer Alley," *LEAN News*, March 1989, p. 1, in LEAN Papers. See also "A Walk for Toxic Justice," *Environmental Action*, January–February 1989, p. 10, clipping in "The Great Louisiana Toxic March" folder, box 5, IUP.

74. Duncan, *Goodbye Green*, 57.

75. "Key Findings: Greenpeace and the Great Louisiana Toxics March," March 8, 1989, "Great Louisiana Toxics March" folder, box 5, IUP.

76. Transcript of *Locked Out!* video, October 7, 1988, pp. 11–24, "BASF v. OCAW Correspondence No. 3" folder, box 12, IUP (quotation on p. 15).

77. "Santa to Confront Grinch in Geismar," December 17, 1988, "BASF Press Releases" folder, box 3, LUP.

Chapter 8. Settlement

1. Miller interview.

2. Marylee M. Orr to Maureen O'Neill, May 3, 1989, "BASF Basagran Plant" folder, LEAN Papers.

3. Ramona Stevens and Marylee Orr to Paul H. Templet, September 22, 1989, "BASF Basagran Plant" folder, LEAN Papers.

4. Marylee M. Orr to Mike McDaniel, May 8, 1989, "1989 LEAN Outgoing" folder, LEAN Papers.

5. "Union Helps Stall BASF Plant Startup," *Chemical and Engineering News*, August 7, 1989, clipping in "Media/Clips/Stats" folder, box 15, IUP.

6. William C. Moran to Maureen O'Neill, April 28, 1989, "BASF Basagran Plant" folder, LEAN Papers.

7. Ibid.

8. Roemer interview.

9. Ibid.; Templet interview; Miller interview.

10. Vicki Ferstel, "BASF Plant Delayed by DEQ Ruling," *Baton Rouge Morning Advocate*, July 8, 1989, clipping in "Media/Clips/Statistics" folder, box 15, IUP.

11. Templet interview; Vicki Ferstel, "BASF Plant Delayed by DEQ Ruling," *Baton Rouge Morning Advocate*, July 8, 1989, clipping in "Media/Clips/Statistics" folder, box 15, IUP; "Union Helps Stall BASF Plant Startup," *Chemical and Engineering News*, August 7, 1989, clipping in "Media/Clips/Stats" folder, box 15, IUP.

12. "The Urgent Lesson of Earth Day," *Chemical Week*, April 29, 1970, p. 8. In other articles in the 1970s, *Chemical Week* stressed that executives must address environmental issues, which represented "the most critical challenge for the process industries in the '70s." "Our Team Tackles the Environment," *Chemical Week*, June 17, 1970, p. 5. This concern continued in the 1980s. See, for example, "Good Public Works Need a Strong Public Voice," *Chemical Week*, June 13, 1984, p. 3.

13. "Why Not Exchange Wastes for Profits?" *Chemical Week*, May 16, 1984, p. 3.

14. "Bridging the Bhopal Gap," *Chemical Week*, May 1, 1985, p. 3; "The Environmental Activists: They've Grown in Competence, and They're Working Together," *Chemical Week*, October 19, 1983, pp. 48, 49, 50, 54, 55.

15. "BASF's Geismar Debacle," *Chemical Week*, August 23, 1989, p. 8; "Rethinking the Backyard Issue," *Chemical Week*, August 23, 1989, p. 5.

16. Donaldson interview; Story interview.

17. Steimel interview.

18. "Employee Communication," August 25, 1987, Story Papers.

19. Les Story, address to Rotary Club, September 24, 1986, Story Papers.

20. "BASF Board of Executive Directors Holds Annual Meeting," *BASF information*, November 1989, p. 1, "BASF Info (U.S.) 1989" folder, box 1, IUP; Frances Frank Marcus, "Labor Dispute in Louisiana Ends with Ecological Gain," *New York*

Times, January 3, 1990, p. A2, clipping in "Labor/End Lockout—Settlement" folder, box 6, IUP.

21. Les Story, address to Rotary Club, September 24, 1986, Story Papers.

22. "Annual Report 1988 BASF Corporation," p. 1, "BASF—Annual Reports—U.S." folder, box 1, IUP; "Stein Reports 1989 Sales, Earnings," *BASF Information*, April 1990, "BASF Info. (U.S.) 1989" folder, box 1, IUP, p. 1.

23. Donaldson interview; John Hall, "Firm Will Build in Geismar," *New Orleans Times-Picayune*, March 30, 1992, p. D2; Pederson, *International Directory*, 51.

24. "BASF Lockout—The Fifth Year," October 5, 1988, "Fifth Year" folder, box 14, IUP; "Five Year Statement of OCAW Local 4-620 on the BASF Lockout," June 15, 1989, "BASF Press Releases" folder, box 3, LUP.

25. Landaiche interview; Joseph M. Misbrener to the Officers and Membership of Local 4-620, December 15, 1989, "BASF Press Releases" folder, box 3, LUP; Rousselle, quoted in "Swedish TV film" videotape, copy in LUP.

26. Gary M. Fink, for example, has described the South as "the Achilles' heel of the American labor movement." Fink, "Fragile Alliance," 784. Many studies have explored the reasons for the weakness of unions in the South. See, for example, Goldfield, "The Failure of Operation Dixie"; Carlton, "The State and the Worker"; Terrill, "No Union for Me." For an overview, see Simon, "Rethinking Why."

27. Henry Kramer, letter to author, February 6, 2001, copy in author's possession.

28. Fink interview; Gremillion interview.

29. "One Thousand March on Third Anniversary of Lockout," OCAW Press Release, n.d., "Corporate Campaign" folder, box 4, LUP; Rousselle interview.

30. For discussion about the centrality of religion to southern culture, see Reed, *The Enduring South*, 63–8, 83–90; Tindall, "Southerners," 13–4. For the reluctance of Protestant churches to support unionization in the textile South, see Pope, *Millhands and Preachers*, 165–9, 201–2; Earle, Knudson, and Shriver, *Spindles and Spires*, 199–206.

31. E. F. Gremillion to the Most Reverend Stanley J. Ott, July 1, 1987, "Religious Letters on BASF Lockout" folder, box 4, LUP.

32. L. J. Story to Reverend Robert Chandler, October 1, 1987, "Religious Letters on BASF Lockout" folder, box 4, LUP; Story interview; Joseph M. Misbrener to the Most Reverend Stanley Joseph Ott, January 16, 1990, "Incoming Correspondence' folder, box 3, LUP.

33. Rosemary M. Collyer to John W. McKendree, February 6, 1989, "BASF and Local 4-620 ULP15–CA-10494 Pleadings No. 2" folder, box 19, IUP.

34. "Five Year Statement of OCAW Local 4-620 on the BASF Lockout," June 15, 1989, "BASF Press Releases" folder, box 3, LUP.

35. For an illustration of union allegations that the NLRB became antiunion during the Reagan presidency, see "The AFL-CIO Condemns the Federal Labor Law, 1985," in *Major Problems*, ed. Boris and Lichtenstein, 592–4.

36. Donaldson interview; John H. Kirkman, letter to author, December 27, 2000, copy in author's possession.

37. Frank Swoboda, "5½-Year Labor Lockout Ends," *Washington Post*, December 16, 1989, clipping in "Labor/End Lockout—Settlement" folder, box 6, IUP; Vicki Ferstel, "BASF, OCAW Reach Tentative Agreement," *Baton Rouge Morning Advocate*, December 15, 1989, clipping in "Labor/End Lockout—Settlement" folder, box 6, IUP; Donaldson interview; "Agreement between BASF Corporation and the Oil, Chemical, and Atomic Workers International Union, AFL-CIO, Local No. 4-620," December 18, 1989, box 6, LUP.

38. Vicki Ferstel, "OCAW Ratifies Three-Year BASF Contract," *Baton Rouge Morning Advocate*, December 19, 1989, clipping in "Labor/End of Lockout—Settlement" folder, box 6, IUP.

39. Donaldson interview; John H. Kirkman, letter to author, December 27, 2000, copy in author's possession.

40. Vicki Ferstel, "Lockout Led to Union's Environmental Activism," *Baton Rouge State-Times*, n.d.; Vicki Ferstel, "OCAW Ratifies Three-Year BASF contract," *Baton Rouge Morning Advocate*, December 19, 1989, clippings in "Labor/End Lockout—Settlement" folder, box 6, IUP.

41. Rand Wilson, "A Labor-Environmental Alliance," *Toxic Times*, summer 1990, pp. 22–3, copy in "Post Lockout" folder, box 11, IUP; Frances Frank Marcus, "Labor Dispute in Louisiana Ends with Ecological Gain," *New York Times*, January 3, 1990, p. A2, clipping in "Labor/End Lockout—Settlement" folder, box 6, IUP.

42. Ramsey Clark to John W. McKendree, January 16, 1990, Stanley Joseph Ott to Esnard Gremillion, December 18, 1989, "Incoming Correspondence" folder, "Richardson" box, LUP.

43. J. L. Watts to John Daigle, January 10, 1990, R. J. Christie to Esnard Gremillion, December 19, 1989, "Incoming Correspondence" folder, box 3, LUP.

44. Resolution on Union Victories, January 5, 1990, "Incoming Correspondence" folder, box 3, LUP; Crowe, quoted in Vicki Ferstel, "Lockout Led to Union's Environmental Activism," *Baton Rouge State-Times*, n.d., clipping in "Labor/End Lockout—Settlement" folder, box 6, IUP.

45. Press Associates Release, January 1, 1990, "Labor/End Lockout—Settlement" folder, box 6, IUP.

46. Frank Swoboda, "5½-Year Labor Lockout Ends," *Washington Post*, December 16, 1989, clipping in "Labor/End Lockout—Settlement" folder, box 6, IUP.

47. "Workers Vote on Plan to End BASF Dispute," *New Orleans Times-Picayune*, December 19, 1989, p. D1.

48. "BASF Lockout—The Fifth Year," October 5, 1988, "Fifth Year" folder, box 14, IUP.

49. Gilin interview.

50. "Summary—Costs to BASF for Lockout," July 1, 1988, "Corporate Campaign Effectiveness" folder, box 2, IUP; Story interview; Jenkins interview.

51. A decertification occurred if more than 30 percent of the workforce petitioned the NLRB for a new election and if a majority of workers then voted against the union.

52. Miller interview. For a graphic example of a strong local union that was decertified after the hiring of permanent replacements, see Getman, *The Betrayal of Local 14*, 192–200. For the rising number of decertifications in the 1980s, see Goldfield, *The Decline of Organized Labor*, 51–2, 159–60.

53. Schneider interview; Fink interview.

54. "Employees not Returning (Retired/Resigned)," "Incoming Correspondence" folder, box 3, LUP.

55. Donaldson interview.

56. Arnold interview.

57. Braud interview.

Chapter 9. No More Trade-Offs

1. "Workers Vote on Plan to End BASF Dispute," *New Orleans Times-Picayune*, December 19, 1989, p. D1.

2. Hawkins interview.

3. Schneider interview.

4. Bobby Schneider testimony, "Labor/End Lockout—Settlement" folder, box 6, IUP.

5. Braud interview; Harvey interview.

6. Arnold interview.

7. Zack Nauth, "Workers Knock Out Chemical Giant," *In These Times*, January 24–30, 1990, pp. 12, 13, 22 (quotation on p. 22).

8. Harvey interview; Fink interview; Smith interview.

9. Darryl Stevens testimony, "Labor/End Lockout—Settlement" folder, box 6, IUP.

10. Fink interview; Smith interview; Landaiche interview; Braud interview; Hawkins interview; Arnold interview.

11. Vicki Ferstel, "BASF, Union Together, but Apart," *Baton Rouge Morning Advocate*, January 20, 1991, p. 6B.

12. "Formal OSHA Complaint Filed by the Oil, Chemical, and Atomic Workers (OCAW) Local 4-620 and Its International Union Concerning Health and Safety Violations at BASF Corp., Geismar, Louisiana," November 1991, "John Daigle" box, LUP.

13. Fink interview.

14. Leonard and Nauth, "Beating BASF," 47.

15. Miller interview; John A. Heilala to David J. Buchner, February 14, 1986, Story Papers; King interview; *The Observer*, July 1991, p. 2, "4-620 Newsletter" folder, box 5, LUP.

16. King interview; Fontenot interview.

17. Duke King et al. to OCAW Local 4-620 Member, March 26, 1990, "Post Lockout" folder, box 11, IUP.

18. "LLNP Summary," January 2000, copy in author's possession; "LLNP Proposal," n.d., copy in author's possession.

19. Grant Proposal for the Louisiana Labor/Neighbor Alliance, October 4, 1991, "John Daigle" box, LUP.

20. Grant Proposal, October 5, 1991, "John Daigle" box, LUP; Rousselle interview; Gremillion interview.

21. Schwab, *Deeper Shades of Green*, 242; "Louisiana 'Watch' Launches Efforts to Protect Workers and Advocate for Safer Workplaces," October 5, 1989, "Outgoing 1989–90 Correspondence" folder, box 5, LUP; King interview.

22. Esnard Gremillion to Chairman, Louisiana Board of Commerce and Industry, June 6, 1989, "Outgoing Correspondence 1989–90" folder, box 5, LUP; "Grant Money Raised by Richard Miller for Projects Spun Off from the BASF Campaign at OCAW Local 4-620," September 13, 1989, "Post Lockout—LCTJ" folder, box 11, IUP.

23. Nauth, *The Great Louisiana Tax Giveaway*, v.

24. *LCTJ News*, December 1992, pp. 1, 7, *LCTJ News*, May 1991, pp. 1, 14, "Post Lockout—LCTJ" folder, box 11, IUP.

25. Leonard and Nauth, "Beating BASF," 44.

26. Schwab, *Deeper Shades of Green*, 248.

27. *LCTJ News*, December 1992, p. 3, "Post Lockout—LCTJ" folder, box 11, IUP; Nauth, *The Great Louisiana Tax Giveaway* (quotation from title).

28. Borne interview; Fontenot interview.

29. Thomas Estabrook, Richard Miller, and Amos Favorite to Maureen O'Neill, May 9, 1991, "John Daigle" box, LUP.

30. *The Observer*, September 1991, p. 5, "4-620 Newsletter" folder, box 5, LUP.

31. *The Observer*, November 1992, p. 4, "4-620 Newsletter" folder, box 5, LUP.

32. Mark Schleifstein, "BASF Cleanup Threatens Water, Critics Say," *New Orleans Times-Picayune*, July 19, 1992, pp. B1, B2; Gilin interview.

33. "Drinking Water Quality in Geismar and Dutchtown," n.d., "Geismar/Dutchtown 1992" folder, "BASF Lockout" box, LUP.

34. Grant Proposal for the Louisiana Labor/Neighbor Alliance, October 4, 1991, "John Daigle" box, LUP.

35. Schwab, *Deeper Shades of Green*, 241–2; A. D. Riley et al. to Neighbor, December 20, 1991, "Geismar/Dutchtown 1992" folder, "BASF Lockout" box, LUP; Grant Proposal for the Louisiana Labor/Neighbor Alliance, October 4, 1991, "John Daigle" box, LUP; *The Observer*, April 1995, p. 7, "4-620 Newsletter" folder, box 5, LUP.

36. Fontenot interview.

37. *The Observer*, September 1993, p. 3, "4-620 Newsletter' folder, box 5, LUP; Carson, *Silent Spring*, 191.

38. *The Observer*, April 1995, p. 2, "4-620 Newsletter" folder, box 5, LUP.

39. Ibid., p. 6.

40. *The Observer*, June 1994, p. 5, "4-620 Newsletters" folder, box 5, LUP; Schneider interview; Arnold interview.

41. "LLNP Summary," January 2000, copy in author's possession; "LLNP Proposal," n.d., copy in author's possession.

42. Ibid; "Concerned African-American residents of Convent and St. James Parish to Carole Browner," May 12, 1997, "Shintech" box, LUP.

43. "From Plantations to Plants: Report of the Emergency National Commission on Environmental and Economic Justice in St. James Parish, Louisiana," n.d., "Shintech" box, LUP (quotations on pp. 27, 32).

44. Albertha Hasten to Nathalie Walker, July 6, 1999, "Shintech WBR/ Iberville PR" folder, "Shintech" box, LUP; Fontenot interview.

45. Hasten interview.

46. Ibid.

47. Orr interview.

48. Borne interview.

49. Bernard Chaillot, "Watchdog Groups Offer Suggestions," unidentified clipping dated October 1, 1991, "LEAN Board Retreat, February 15–16, 1992" folder, LEAN Papers.

50. Bob Anderson, "Many Take Credit for Reduced Toxic Releases," unidentified and undated clipping in "LEAN Board Retreat, February 15–16, 1992" folder, LEAN Papers; "Environmental Achievements Growing at BASF Geismar," *BASF Community Report*, December 1995, pp. 1, 3, "BASF" file, Ascension Parish Library; Favorite interview.

51. Richard Miller to Nicola W. Palmieri-Egger, April 28, 1990, "Outgoing 1989–90 Correspondence" folder, box 5, LUP.

52. *The Observer*, September 1993, p. 3, "4-620 Newsletter" folder, box 5, LUP.

53. Donaldson interview; Harvey interview; Schneider interview.

54. Harvey interview; Schneider interview.

55. Schneider interview; Rousselle interview.

56. Schneider interview; King interview.

57. Schneider interview.

58. "Hotline to Open on Area Chemical Plants," *Baton Rouge Morning Advocate*, clipping dated June 26, 1992, "Chemical Industry" file, Ascension Parish Library.

59. "Environmental Achievements Growing at Geismar," *BASF Community Report*, December 1995, pp. 1, 3, "COE Students Learn by Doing," *BASF Community Report*, June 1995, p. 1, "BASF Awards College Scholarships," BASF Community Report, fall 1992, p. 1, "BASF" folder, Ascension Parish Library.

60. "Meet Phil Greeson: BASF Geismar Site's New General Manager Seeking Active Role in Community Service," *BASF Community Report*, December 1995, p. 2, in "BASF" file, Ascension Parish Library.

61. "Industry Report," *Greater Baton Rouge Business Report*, January 28, 1992, clipping in "LEAN Board Retreat, February 15–16, 1992" folder, LEAN Papers.

62. "Partners in Louisiana: Working for Louisiana," March 4, 1997, p. B5, "Chemical Industry" file, Ascension Parish Library.

63. Badger, "When I Took the Oath," 16, 17, 20; Schwab, *Deeper Shades of Green*, 278–80.

64. Templet interview; Barbara Koppel, "Cancer Alley, Louisiana," *Nation*, November 8, 1999, pp. 20, 22, 24.

65. "From Plantations to Plants: Report of the Emergency National Commission on Environmental and Economic Justice in St. James Parish," n.d., pp. 1, 13–14, "Shintech" box, LUP (quotation on p. 1); Paul Templet, "Industrial Development in Louisiana: Shintech Case Study," March 1997, p. 3, copy of research paper in author's possession.

66. Duncan, *Goodbye Green*, 32.

67. Barbara Koppel, "Cancer Alley, Louisiana," *Nation*, November 8, 1999, pp. 16–20, 22, 24; Borne interview; "Partners in Louisiana: Working for Louisiana," March 4, 1997, pp. B34–36, "Chemical Industry" file, Ascension Parish Library.

68. Barbara Koppel, "Cancer Alley, Louisiana," *Nation*, November 8, 1999, pp. 16–20, 22, 24 (quotation on p. 19). See also "'Cancer Alley' Not Real, La. Researcher Reports," *Baton Rouge Morning Advocate*, October 3, 2000, p. 15A.

69. Orr interview.

Conclusion

1. For the concentration on decline, see, for example, Goldfield, *The Decline of Organized Labor;* Moody, *An Injury to All;* Geoghegan, *Which Side Are You On?;* Nissen, ed., *U.S. Labor Relations;* Davis, *Prisoners of the American Dream*, esp. 102–53; Edsall, *The New Politics of Inequality*, 141–78. Zieger, *American Workers*, 193–205, and Dark, *The Unions and the Democrats*, esp. 1–2, 13–21, provide useful overviews.

2. Dark, *The Unions and the Democrats*, 15.

3. Zieger, *American Workers*, 193; Dubofsky, "Jimmy Carter," 96–7.

4. John Lawrence, quoted in "ABC 20/20" videotape, n.d., box 8, IUP, copy in author's possession.

5. Geoghegan, *Which Side Are You On?*, 3, 232.

6. Zieger, *American Workers*, 199; Juravich and Bronfenbrenner, *Ravenswood*, 202–5 (quotation from book title).

7. Minchin, "Permanent Replacements."

8. Dark, *The Unions and the Democrats* (quotation from the title of part 1); Shostak, *Robust Unionism*, 258–9; Zieger, *American Workers*, 204–5; Craver, *Can Unions Survive?*, 156–7; Juravich and Bronfenbrenner, *Ravenswood*, 205–6.

9. Matt Witt, "Labor's New Leverage: Unions Are Forging Alliances That Corporations May Regret," *Washington Post*, May 10, 1987, clipping in "Clips/Media/Statistics" folder, box 15, IUP; Stefan Schaaf, "In den USA wird der Arbeitskampf zur PR-Schlacht," *Die Weltwoche*, July 2, 1987, p. 21, clipping in "1987" folder, untitled box, LUP.

10. "Union Militancy—and Beyond," *Barron's*, June 8, 1987, clipping in "Clips/Media/Statistics" folder, box 15, IUP; "Corporate Campaign Replaces Strike as Primary Union Tool," *Business International*, September 22, 1986, clipping in "Media/Clips/Stats" folder, box 15, IUP.

11. Jonathan Tasini, "Labor Unions Fine-Tune Battle," *New York Newsday*, December 31, 1989, clipping in "BASF Press Releases" folder, box 3, LUP.

12. Juravich and Bronfenbrenner, *Ravenswood*, 203. The International Paper strike ended with many strikers complaining about a lack of support from the international union. See, for example, Getman, *The Betrayal of Local 14*, esp. 62–4, 199–200; Martin interview; Wilt interview.

13. Leonard interview.

14. Bullard, *Dumping in Dixie*, xiii–xvi, 50–4, 69–73, 103–7; McCaull, "Discriminatory Air Pollution"; Reynolds, "Triana, Alabama"; Hays, *A History of Environmental Politics*, 60.

15. Schwab, *Deeper Shades of Green*, xvii–xxii; Gottlieb, *Forcing the Spring*, 3–11; Taylor, "Do Environmentalists Care about Poor People?"; Gale, "The Environmental Movement and the Left."

16. OCAW president Joseph M. Misbrener, for example, described the lockout as "a model to all of us in the trade union movement." Joseph M. Misbrener to Local union presidents and secretary-treasurers, December 15, 1989, "Incoming Correspondence" folder, "Richardson" box, LUP.

17. Statistics supplied by PACE International Union, copy in author's possession.

18. Wages interview.

19. Mazzochi interview; Leonard interview.

20. Statistics supplied by PACE International Union, copy in author's possession; Wages interview.

21. Leonard and Nauth, *Beating BASF*, 48–9 (quotation on p. 49); *Out of Control*, videotape, box 8, IUP, copy in author's possession.

22. "'Technical'?: Workers Complaints Validated," *Charleston Gazette*, November 30, 1994, p. 4A; David White to File, October 10, 1994, "Enviro-Air" folder, "Oxy No. 1" box, IUP; "Oxychem Challenges Accuracy of OCAW's Environmental Allegations," Oxychem Press Release dated November 23, 1994, "Correspondence from Oxy" folder, "Oxy No. 1" box, IUP.

23. "Dirty Business: A Response to Crown Central Petroleum's Campaign of Misinformation," Texans United Education Fund and the Oil, Chemical, and Atomic Workers Union, November 1997, copy in author's possession; "More Dirty Business: Crown Petroleum in Trouble Again," Texans United Education Fund and Paper, Allied-Industrial, Chemical, and Energy Workers International Union, April 2000, copy in author's possession.

24. Dewey, "Working for the Environment," 45, 51–3 (quotation on p. 45).

25. Miller, "Towards an Environmental/Labor Coalition," 33–6; Temkin,

"State, Ecology, and Independence," 451–4; Temkin, "The Political Impact," 182–231, 323–51.

26. Gordon, "Shell No!"; Shostak, *Robust Unionism*, 266–71; Sackman, "Nature's Workshop," 29. For a successful effort at integrating economic and environmental history, see Cronon, *Nature's Metropolis*.

27. Gottlieb, *Forcing the Spring*, 304.

28. Gordon, "Shell No!" 482; Schwab, *Deeper Shades of Green*, xiv, 234–50.

29. *In These Times*, February 21–27, 1990, p. 15, clipping in "Post Lockout" folder, box 11, IUP.

30. Montrie, "Expedient Environmentalism."

31. Malek-Wiley interview.

32. Orr interview; Gottlieb, *Forcing the Spring*, 298; "Brothers and Sisters Greens and Labor: It's a Coalition That Gives Corporate Polluter Fits," *Sierra*, January–February 1999, printed off http://www.webshells.com/crown/texts/moberg.htm; Mazzochi interview.

33. Gottlieb, *Forcing the Spring*, 300–1; "Brothers and Sisters Greens and Labor: It's a Coalition That Gives Corporate Polluter Fits," *Sierra*, January-February 1999, printed off http://www.webshells.com/crown/texts/moberg.htm.

34. Story interview; Jenkins interview; Donaldson interview.

35. Leonard interview; Miller interview.

36. *LCTJ News*, December 1992, p. 3, in "Post Lockout—LCTJ" folder, box 11, IUP; Gremillion interview.

37. "Friend and Brother Forever" memorial, in Local 4-620's union hall, Gonzales, La.; *The Observer*, June 1994, p. 1, "4-620 Newsletter" folder, box 5, LUP; John Daigle, "Comments on Ciba Geigy's Hazardous Waste Incinerator Permit Application, St. Gabriel," May 24, 1989, "John Daigle" box, LUP.

38. Favorite interview.

Bibliography

Anthony, Carl. "Why African-Americans Should be Environmentalists." In *Major Problems in American Environmental History*, ed. Carolyn Merchant, 540–43. Lexington, Mass.: D. C. Heath, 1993.

Anthony, Richard. "Polls, Pollution, and Politics: Trends in Public Opinion on the Environment." *Environment* 24:4 (May 1982): 14–20, 33–34.

Arnold, Roger. Interview with author. Gonzales, La., August 8, 2000.

Badger, Tony. "'When I Took the Oath of Office, I Took No Vow of Poverty': Race, Corruption, and Democracy in Louisiana, 1928–2000." Unpublished paper in author's possession.

Bartley, Numan V. *The New South: The Story of the South's Modernization, 1945–1980*. Baton Rouge: Louisiana State University Press, 1995.

Berghahn, V. R. *Modern Germany: Society, Economy, and Politics in the Twentieth Century*. 2nd ed. Cambridge, England: Cambridge University Press, 1987.

Berghahn, Volker R., and Detlev Karsten. *Industrial Relations in West Germany*. Oxford, England: Berg, 1987.

Berman, William C. *America's Right Turn: From Nixon to Bush*. Baltimore: Johns Hopkins University Press, 1994.

Boris, Eileen, and Nelson Lichtenstein, eds. *Major Problems in the History of American Workers*. Lexington, Mass.: D. C. Heath, 1991.

Borkin, Joseph. *The Crime and Punishment of I.G. Farben*. London: Andre Deutsch, 1979.

Borne, Dan. Interview with author. Baton Rouge, September 13, 2000.

Boyer, Paul S., Clifford E. Clark, Jr., Sandra McNair Hawley, Joseph F. Kett, Neal Salisbury, Harvard Sitkoff, and Nancy Woloch. *The Enduring Vision: A History of the American People*. Concise 3rd ed. Boston: Houghton Mifflin, 1998.

Brattain, Michelle. *The Politics of Whiteness: Race, Workers, and Culture in the Modern South*. Princeton: Princeton University Press, 2001.

Braud, Marion "Putsy." Interview with author. Gonzales, La., August 17, 2000.

Brody, David. "The Breakdown of Labor's Social Contract." *Dissent* (winter 1992): 32–41.

Brown, Michael H. *Laying Waste: The Poisoning Of America by Toxic Chemicals*. New York: Pantheon Books, 1980.

Budiansky, Stephen. *Nature's Keepers: The New Science of Nature Management*. New York: Free Press, 1995.

Bullard, Robert D. *Dumping in Dixie: Race, Class, and Environmental Quality*. Boulder, Colo.: Westview Press, 1990.

Carlton, David L. "The State and the Worker in the South: A Lesson from South Carolina." In *The Meaning of South Carolina History: Essays in Honor of George C. Rogers, Jr.*, ed. David R. Chesnutt and Clyde N. Wilson, 186–201. Columbia: University of South Carolina Press, 1991.

Carson, Rachel. *Silent Spring*. London: Hamish Hamilton, 1962.

Colten, Craig E. "Texas v. the Petrochemical Industry: Contesting Pollution in an Era of Industrial Growth." In *The Second Wave: Southern Industrialization from the 1940s to the 1970s*, ed. Philip Scranton. Athens: University of Georgia Press, 2001.

Cowdrey, Albert E. *This Land, This South: An Environmental History*. Lexington: University Press of Kentucky, 1983.

Craver, Charles B. *Can Unions Survive?: The Rejuvenation of the American Labor Movement*. New York: New York University Press, 1993.

Cronon, William. *Nature's Metropolis: Chicago and the Great West*. New York: W. W. Norton, 1991.

Dark, Taylor E. *The Unions and the Democrats: An Enduring Alliance*. Ithaca, N.Y.: ILR Press, 1999.

Davidson, Ray. *Challenging the Giants: A History of the Oil, Chemical, and Atomic Workers International Union*. Denver: Oil, Chemical, and Atomic Workers International Union, 1988.

———. *Peril on the Job: A Study of Hazards in the Chemical Industries*. Washington, D.C.: Public Affairs Press, 1970.

Davis, Mike. *Prisoners of the American Dream: Politics and Economy in the History of the U.S. Working Class*. London: Verso, 1986.

Dedeaux, Myron. Interview with author. Gonzales, La., September 7, 2000.

Derr, Mark. *Some Kind of Paradise: A Chronicle of Man and the Land in Florida*. Gainesville: University Press of Florida, 1998.

Dewey, Scott. "Working for the Environment: Organized Labor and the Origins of Environmentalism in the United States, 1948–1970." *Environmental History*, 3:1 (January 1998): 45–63.

Donaldson, Richard. Interview with author. Baton Rouge, November 7, 2000.

———. Correspondence with author. January 20, 2001, and February 9, 2001. Copies in author's possession.

Draper, Alan. *Conflict of Interests: Organized Labor and the Civil Rights Movement in the South, 1954–1968*. Ithaca, N.Y.: ILR Press, 1994.

Dubofsky, Melvyn. "Jimmy Carter and the End of the Politics of Productivity." In *The Carter Presidency: Policy Choices in the Post-New Deal Era*, ed. Gary M. Fink and Hugh Davis Graham, 95–116. Lawrence: University Press of Kansas, 1998.

———. *The State and Labor in Modern America.* Chapel Hill: University of North Carolina Press, 1994.

Duncan, Glen A. *Goodbye Green: How Extremists Stole the Environmental Movement from Middle America and Killed It.* Bellevue, Wash.: Merril Press, 2000.

Dunlap, Riley E. "Polls, Pollution, and Politics Revisited: Public Opinion on the Environment in the Reagan Era." *Environment* 29:6 (July-August 1987): 6–11, 32–7.

Earle, John R., Dean D. Knudson, and Donald W. Shriver, Jr. *Spindles and Spires: A Restudy of Religion and Social Change in Gastonia.* Atlanta: John Knox Press, 1976.

Edsall, Thomas Byrne. *The New Politics of Inequality.* New York: W. W. Norton, 1984.

Epstein, Samuel S., Lester O. Brown, and Carl Pope. *Hazardous Waste in America.* San Francisco: Sierra Club Books, 1982.

Everest, Larry. *Behind the Poison Cloud: Union Carbide's Bhopal Massacre.* Chicago: Banner Press, 1986.

Fairclough, Adam. *Race and Democracy: The Civil Rights Struggle in Louisiana, 1915–1972.* Athens, Ga.: University of Georgia Press, 1995.

Favorite, Amos. Interview with author. Geismar, La., August 10, 2000.

Fink, Gary M. "Fragile Alliance: Jimmy Carter and the American Labor Movement." In *The Presidency and Domestic Policies of Jimmy Carter,* ed. Herbert D. Rosenbaum and Alexej Ugrinsky, 783–803. Westport, Conn.: Greenwood Press, 1994.

Fink, Gary M., and Merl E. Reed, eds. *Race, Class, and Community in Southern Labor History.* Tuscaloosa: University of Alabama Press, 1994.

Fink, Roy. Interview with author. Gonzales, La., August 15, 2000.

Fontenot, Willie. Interview with author. Baton Rouge, August 10, 2000.

Gale, Richard P. "The Environmental Movement and the Left: Antagonists or Allies?" *Sociological Inquiry* 53 (spring 1983): 179–99.

Garrow, David J., *Bearing the Cross: Martin Luther King, Jr., and the Southern Christian Leadership Conference.* 2nd ed. London: Vintage, 1993.

———. *Protest at Selma: Martin Luther King, Jr., and the Voting Rights Act of 1965.* New Haven, Conn.: Yale University Press, 1978.

Geoghegan, Thomas. *Which Side Are You On?: Trying to Be for Labor When It's Flat on Its Back.* New York: Farrar, Straus, and Giroux, 1991.

Getman, Julius. *The Betrayal of Local 14: Paperworkers, Politics, and Permanent Replacements.* Ithaca, N.Y.: ILR Press, 1998.

Gilin, David. Interview with author. Baton Rouge, September 13, 2000.

Goldfarb, Theodore. "Environmental Legislation." In *Major Problems in American Environmental History,* ed. Carolyn Merchant, 547–51. Lexington, Mass.: D. C. Heath, 1993.

Goldfield, Michael. *The Decline of Organized Labor in the United States.* Chicago: University of Chicago Press, 1987.

————. "The Failure of Operation Dixie: A Critical Turning Point in American Political Development?" In *Race, Class, and Community in Southern Labor History*, ed. Gary M. Fink and Merl E. Reed, 166–89. Tuscaloosa: University of Alabama Press, 1994.

Gordon, Robert. "'Shell No!' OCAW and the Labor-Environmental Alliance." *Environmental History* 3:4 (October 1998): 460–87.

Gottlieb, Robert. *Forcing the Spring: The Transformation of the American Environmental Movement*. Washington, D.C.: Island Press, 1993.

Graham, Frank, Jr. *Since Silent Spring*. Greenwich, Conn.: Fawcett Crest, 1970.

Gremillion, Esnard. Interview with author. Gonzales, La., August 7, 2000.

Guidry, Dexter. Interview with author. Gonzales, La., August 31, 2000.

Guste, William. Interview with author. Baton Rouge, September 14, 2000.

Harvey, Gladys. Interview with author. Gonzales, La., August 21, 2000.

Hasten, Albertha. Interview with author. Gonzales, La., August 23, 2000.

Hawkins, Carey. Interview with author. Gonzales, La., August 16, 2000.

Hays, Samuel P. *Beauty, Health, and Permanence: Environmental Politics in the United States, 1955–1985*. Cambridge, England: Cambridge University Press, 1987.

————. *A History of Environmental Politics since 1945*. Pittsburgh: University of Pittsburgh Press, 2000.

Hodges, James A. "J. P. Stevens and the Union: Struggle for the South." In *Race, Class, and Community in Southern Labor History*, ed. Gary M. Fink and Merl E. Reed, 53–64. Tuscaloosa: University of Alabama Press, 1994.

————. "The Real Norma Rae." In *Southern Labor in Transition, 1940–1995*, ed. Robert H. Zieger, 251–72. Knoxville: University of Tennessee Press, 1997.

Hoerr, John. "Solidaritas at Harvard: Meet the Harvard of the Labor Movement, a Model of the New Unionism." *American Prospect* 14 (summer 1993): 67–82.

Hulsberg, Werner. *The German Greens: A Social and Political Profile*. London: Verso, 1988.

Jenkins, Bill. Interview with author. Knoxville, Tenn., November 5 and 6, 2000.

Juravich, Tom, and Kate Bronfenbrenner. *Ravenswood: The Steelworkers' Victory and the Revival of American Labor*. Ithaca, N.Y.: Cornell University Press, 1999.

Kazis, Richard, and Richard L. Grossman. *Fear at Work: Job Blackmail, Labor, and the Environment*. New York: Pilgrim Press, 1982.

Kettenacker, Lothar. *Germany since 1945*. Oxford, England: Oxford University Press, 1997.

Kilborn, Peter T. "Replacement Workers: Management's Big Gun." In *Major Problems in the History of American Workers*, ed. Eileen Boris and Nelson Lichtenstein, 598–600. Lexington, Mass.: D. C. Heath, 1991.

King, Noel "Duke." Telephone interview with author. July 31, 2001.

Kirkman, John. Correspondence with author. December 27, 2000, February 2, 2001, July 19, 2001. Copies in author's possession.

Kramer, Alan. *The West German Economy, 1945–1955*. Oxford, England: Berg, 1991.

Kramer, Henry. Correspondence with author. December 11, 2000, January 22, 2001, January 24, 2001, February 6, 2001, February 7, 2001, February 9, 2001. Copies in author's possession.

Landaiche, Tommy. Interview with author. Gonzales, La., August 9, 2000.

Leonard, Richard. Interview with author. Nashville, July 2, 2000.

Leonard, Richard, and Zack Nauth. "Beating BASF: OCAW Busts Union-Buster." *Labor Research Review* 16 (1991): 34–49.

Locked Out! Produced by Christopher Bedford. Washington, D.C.: Organizing Media Project, 1988. Videotape.

Louisiana Environmental Action Network Papers. Louisiana Environmental Action Network, Baton Rouge.

McCaull, Julian. "Discriminatory Air Pollution: If Poor, Don't Breathe." *Environment* 18:2 (1976): 26–31.

Malek-Wiley, Darryl. Interview with author. New Orleans, August 16, 2000.

Marchand, Sidney A. *The Story of Ascension Parish, Louisiana.* Donaldsonville, La.: Sidney A. Marchand, 1931.

Martin, Dale. Interview with author. Lock Haven, Pa., July 17, 1998.

Mazzochi, Tony. Interview with author. Washington, D.C., September 8, 2000.

Merchant, Carolyn, ed. *Major Problems in American Environmental History.* Lexington, Mass.: D. C. Heath, 1993.

Miller, Alan S. "Towards an Environmental/Labor Coalition," *Environment* 22:5 (1980): 32–9.

Miller, Richard. Interview with author. Washington, D.C., September 9, 2000.

———. Personal papers. Holyoke, Mass.

Minchin, Timothy J. "Federal Policy and the Racial Integration of Southern Industry, 1961–1980." *Journal of Policy History,* 11:2 (1999): 147–78.

———. "Permanent Replacements and the Breakdown of the 'Social Accord' in Calera, Alabama, 1974–1999." *Labor History* 42:4 (2001): 371–96.

Montrie, Chad. "Expedient Environmentalism: Opposition to Coal Surface Mining in Appalachia and the United Mine Workers of America, 1945–1975." *Environmental History,* 5:1 (January 2000): 75–98.

Moody, Kim. *An Injury to All: The Decline of American Unionism.* New York: Verso, 1988.

Nader, Ralph, Ronald Brownstein, and John Richard. *Who's Poisoning America: Corporate Polluters and Their Victims in the Chemical Age.* San Francisco: Sierra Club Books, 1981.

Nauth, Zack. *The Great Louisiana Tax Giveaway: A Decade of Corporate Welfare, 1980–1989.* Baton Rouge: Louisiana Coalition for Tax Justice, 1992.

Nissen, Bruce. "A Post–World War II 'Social Accord?'" In *U.S. Labor Relations, 1945–1989: Accommodation and Conflict,* ed. Bruce Nissen, 173–208. New York: Garland Publishing, 1990.

Nordstrom, Laura. Interview with author. Gonzales, La., August 9, 2000.

Norton, Mary Beth, David M. Katzman, Paul D. Escott, Howard P. Chudacoff,

Thomas G. Paterson, William M. Tuttle, Jr., and William J. Brophy. *A People and a Nation: A History of the United States*. 4th brief ed. Boston: Houghton Mifflin, 1996.

OCAW (Oil, Chemical, and Atomic Workers' International Union) Papers. PACE International Union, Nashville.

OCAW (Oil, Chemical, and Atomic Workers' International Union), Local 4-620 Papers. PACE Local 4-620 Union Hall, Gonzales, La.

Orr, Marylee. Interview with author. Baton Rouge, August 30, 2000.

Out of Control. Written, produced, and directed by Christopher Bedford. Washington, D.C.: Organizing Media Project, 1990. Videotape.

Palmieri-Egger, Nicola. Correspondence with author. January 12, 2001. Copy in author's possession.

Patterson, James T. *The Dread Disease: Cancer and Modern American Culture*. Cambridge, Mass.: Harvard University Press, 1987.

Pederson, Jay P., ed. *International Directory of Company Histories*. Vol. 18. New York: St. James Press, 1997.

Pope, Liston. *Millhands and Preachers: A Study of Gastonia*. New Haven, Conn.: Yale University Press, 1942.

Reed, John Shelton. *The Enduring South: Subcultural Persistence in Mass Society*. Lexington, Mass.: D. C. Heath, 1972.

Reynolds, Barbara. "Triana, Alabama: The Unhealthiest Town in America?" *National Wildlife* 18 (August 1980): 33.

Roemer, Charles "Buddy." Interview with author. Baton Rouge, September 12, 2000.

Rousselle, Ernie. Interview with author. Gonzales, La., August 9, 2000.

Sackman, Douglas C. "'Nature's Workshop: The Work Environment and Workers' Bodies in California's Citrus Industry, 1900–1940." *Environmental History* 5:1 (January 2000): 27–53.

Salmond, John A. *"My Mind Set on Freedom": A History of the Civil Rights Movement, 1954–1968*. Chicago: Ivan R. Dee, 1997.

Schneider, Bobby. Interview with author. Galvez, La., August 11, 2000.

Schulman, Bruce J. *From Cotton Belt to Sunbelt: Federal Policy, Economic Development, and the Transformation of the South, 1938–1980*. New York: Oxford University Press, 1991.

Schwab, Jim. *Deeper Shades of Green: The Rise of Blue-Collar and Minority Environmentalism in America*. San Francisco: Sierra Club Books, 1994.

Scranton, Philip, ed. *The Second Wave: Southern Industrialization from the 1940s to the 1970s*. Athens: University of Georgia Press, 2001.

Shostak, Arthur B. *Robust Unionism: Innovations in the Labor Movement*. Ithaca, N.Y.: ILR Press, 1991.

Simon, Bryant. "Rethinking Why There Are So Few Unions in the South." *Georgia Historical Quarterly* 81:2 (summer 1997): 465–84.

Smith, Frank. Interview with author. Gonzales, La., August 9, 2000.

Steimel, Ed. Interview with author. Baton Rouge, September 12, 2000.

Story, Les. Interview with author. Atlanta, November 9, 2000.

———. Personal papers.

Taylor, Dorceta E. "Blacks and the Environment: Toward an Explanation of the Concern and Action Gap between Blacks and Whites." *Environment and Behavior* 21:2 (March 1989): 175–205.

Taylor, Ronald A. "Do Environmentalists Care about Poor People?" *U.S. News and World Report* 96 (April 2, 1984): 51–2.

Temkin, Benny. "The Political Impact of the Energy Crisis on American Working-Class Organizations." Ph.D. diss., Columbia University, 1980.

———. "State, Ecology, and Independence: Responses to the Energy Crisis in the United States." *British Journal of Political Science* 13:4 (October 1983): 441–62.

Templet, Paul. Interview with author. Baton Rouge, September 11, 2000.

Terrill, Thomas E. "'No Union for Me': Southern Textile Workers and Organized Labor." In *The Meaning of South Carolina History: Essays in Honor of George C. Rogers, Jr.*, ed. David R. Chesnutt and Clyde N. Wilson, 202–13. Columbia: University of South Carolina Press, 1991.

Tindall, George B. "Beyond the Mainstream: The Ethnic Southerners." *Journal of Southern History* 15:1 (February 1974): 3–18.

United Paperworkers' International Union Papers. PACE International Union, Nashville.

Vig, Norman J., and Michael E. Kraft, eds. *Environmental Policy in the 1980s: Reagan's New Agenda*. Washington, D.C.: Congressional Quarterly Press, 1984.

Wages, Robert. Interview with author. Nashville, July 3, 2000.

Williams, Juan. *Eyes on the Prize: America's Civil Rights Years, 1954–1965*. New York: Penguin, 1987.

Wilt, Charles. Interview with author. Lock Haven, Pa., July 18, 1998.

Wiltz, Rebecca. Interview with author. Gonzales, La., August 9, 2000.

Zieger, Robert H. *American Workers, American Unions*. 2nd ed. Baltimore: Johns Hopkins University Press, 1994.

Index

Timothy Minchin teaches American history at the University of St. Andrews in Scotland. His previous books include *Hiring the Black Worker: The Racial Integration of the Southern Textile Industry, 1960–1980* (1999), which won the Richard A. Lester Prize, awarded by Princeton University for the best book on industrial relations and labor economics published in 1999. His most recent work is *The Color of Work: The Struggle for Civil Rights in the Southern Paper Industry, 1945–1980* (2001).